BIRDS FROM COAHUILA, MEXICO

EMIL K. URBAN

ZINC READ

Published by
This edition copyright © 2024 by Zinc Read
Design and setting By Zinc Read
Email- zincread@gmail.com
ISBN-978-93-6277-089-9

CONTENTS

INTRODUCTION

The following account is a summary of the present knowledge of the birds of Coahuila. Some 500 specimens from Coahuila in the Museum of Natural History at the University of Kansas are the basis for this report; these are supplemented by records of birds previously listed from the State.

In Coahuila, habitats vary from those characteristic near tree-line to those of the floors of the low deserts. Because of the variety of habitats, many kinds of birds are present in the State; at least 312 living named kinds of 249 species have been recorded. Possibly another 100 species will be reported after further studies have been made there. At least 154 of the species listed in this paper probably breed in Coahuila. The bird fauna in the State includes species characteristic of eastern North America and of western North America, species that range from the Atlantic to the Pacific Ocean, and species found only, or mostly, in Mexico.

I thank Professor E. Raymond Hall, Doctor Richard F. Johnston and Doctor Robert M. Mengel for their kind help, and Doctor Harrison B. Tordoff for first suggesting this study to me. Unless otherwise stated, the nomenclature in this paper is that of the A.O.U. Check-list Committee (1957). Catalogue numbers are those of the Museum of Natural History at the University of Kansas. In so far as known to me, all birds recorded in the literature from Coahuila are listed below. In a few instances the only support for occurrence is the ascription of a given kind to Coahuila (without mention of date, catalogue number, or precise locality) by Friedmann, Griscom, and Moore (1950), and/or the A.O.U. Check-list Committee (1957); when this is so the entire entry is inclosed within brackets. In the accounts beyond, an asterisk indicates that the kind breeds in Coahuila; two asterisks indicate probable breeding in the State.

DISTRIBUTION OF THE KNOWN BREEDING BIRDS OF COAHUILA

Topography and Climate

Coahuila lies in the broad northern end of Mexico, immediately east of the center of the continental mass. The mountains of Coahuila, which are part of the Rocky Mountain-Sierra Madre Oriental Axis, extend in a north-south direction and divide the lower lands into two areas, a larger one, a part of the Central Plateau, to the westward and a smaller one, a part of the Gulf Coastal Plain, to the northeastward. Most of the mountains of Coahuila do not exceed 6000 feet in elevation. A few peaks such as in the Sierra del Carmen, Sierra del Pino, Sierra de la Madera, Sierra Encarnacion, and Sierra de Guadalupe, are more than 9000 feet high, and some more than 10,000 feet in elevation occur near the southeastern border of the State in the Sierra Madre Oriental. The Gulf Coastal Plain of northeastern Coahuila ranges from 700 feet to 1800 feet. The desert plains of the Mesa del Norte to the west of the Sierra Madre Oriental Axis are higher, more rugged, and more dissected than those of the Coastal Plain and are marked by scattered desert ranges, buttes, low hills, and knobs.

Most of Coahuila is arid. Rainfall is moderate on the Coastal Plain and is low west of the central mountains. Baker (1956:128-132) and Muller (1947:35-38) give good summary discussions of the topography and climate of Coahuila, and the reader is referred to these for further details.

Biotic Communities

Baker (1956:132) stated that "the biotic communities of Coahuila might be divided in accordance with the three physiographic areas of the State: the Gulf Coastal Plain, the mountains, and the desert plains of the Mesa del Norte." Goldman and Moore (1945:348-349) listed three biotic provinces in Coahuila: the Chihuahua-Zacatecas Biotic Province, in the western half of the State; the Tamaulipas Biotic Province, in the northeastern part of the State; and the Sierra Madre Oriental Biotic Province, in the southeastern part of the State. Merriam (1898) noted that definable portions of the Lower Sonoran Life-zone, the Upper Sonoran Life-zone, the Transition Life-zone, and the Canadian Life-zone can be distinguished in Coahuila. In my study of the distribution of the avifauna of Coahuila, I found that the three biotic provinces listed by

Goldman and Moore (*op. cit.*) as major headings and Merriam's life-zones as supplements are the most satisfactory divisions.

The Tamaulipas Biotic Province.--This province consists of lowland plains and a few isolated ranges of low mountains. The average rainfall is 23 inches (Baker, 1956:130), considerably more than the 10 inches falling in the western part of the State. In the northeastern section of the State, the moderate amount of rain, mesic vegetation, and close proximity to the eastern migration pathway importantly influence the types of birds found.

In Coahuila, the Coastal Plain and the Rio Grande Plain lie in the path of the northernmost trade winds; they account for the more humid eastern slopes of the mountains of the northeastern part of the State (Muller, 1947:38). Nevertheless, the northeastern section of the State is semi-arid and can be placed in the Lower Sonoran Life-zone. The vegetation consists mainly of thorny shrubs and small trees with a liberal admixture of yuccas, agaves, and cacti, and closely resembles that of southern Texas, northern Nuevo Leon, and northern Tamaulipas (Goldman and Moore, 1945:354).

Migrant birds from the eastern flyway and less commonly migrants from western North America pass through northeastern Coahuila. The following breeding birds seem to be associated with this province: Harris' Hawk, Bobwhite (*C. v. texanus*), Scaled Quail (*C. s. castanogastris*), Yellow-billed Cuckoo, Groove-billed Ani, Green Kingfisher, Golden-fronted Woodpecker, Hairy Woodpecker (*D. v. intermedius*), Ladder-backed Woodpecker (*D. s. symplectus*), Vermilion Flycatcher (*P. r. mexicanus*), Cave Swallow, Gray-breasted Martin, Black-crested Titmouse (*P. a. atricristatus*), Carolina Wren, Long-billed Thrasher, Curve-billed Thrasher (*T. c. oberholseri*), Blue-gray Gnatcatcher (*P. c. caerulea*), Hutton's Vireo (*V. h. carolinae*), Bell's Vireo (*V. b. medius*), Yellowthroated Vireo, Red-eyed Vireo, Summer Tanager (*P. r. rubra*), Olive Sparrow, Cassin's Sparrow, and Black-throated Sparrow (*A. b. bilineata*).

The Sierra Madre Oriental Biotic Province.--Southeastern Coahuila is in this province that includes mountains in southern Nuevo Leon, southwestern Tamaulipas, and eastern San Luis Potosi. Areas classifiable as Canadian, Transition, Upper Sonoran, and Lower Sonoran in life-zone are found in this province. This region of Coahuila receives the highest rainfall; this is evidenced by the luxuriant growth of boreal plants living in the higher places there (Baker, 1956:131). Spruce, pine, and aspen occur at higher elevations and oaks, thorny shrubs, and grasslands are present lower down.

Birds of central or southern Mexico reach the southern part of Coahuila; the Thick-billed Parrot, Hooded Yellowthroat, and Rufous-capped Atlapetes are examples. A boreal forest on the higher slopes of the mountains of southeastern Coahuila is suitable for certain northern birds such as Goshawks, Pine Siskins, and Brown Creepers. Some species of birds ordinarily associated with western North America are present in Coahuila only in its southeastern part; striking examples of disjunction in range thus occur. Probably sometime in the past these birds were distributed throughout most of Coahuila. When this area became arid, these species disappeared from all of Coahuila except from the high mountains in the southeastern part. For example, Steller's Jay and the Scrub Jay are absent in the Sierra del Carmen of northwestern Coahuila but do occur in southeastern Coahuila.

Migrants of the eastern flyway as well as migrants associated with western North America pass through this section of Coahuila. The following breeding birds are associated with this province: Goshawk, Band-tailed Pigeon, Thick-billed Parrot, Golden-fronted Woodpecker, Ladder-backed Woodpecker (*D. s. giraudi*), Pine Flycatcher, Buff-breasted Flycatcher, Vermilion Flycatcher (*P. r. mexicanus*), Steller's Jay, Scrub Jay, Mexican Chickadee, Black-crested Titmouse (*P. a. atricristatus*), Cactus Wren (*C. b. guttatus*), Robin, Blue-gray Gnatcatcher (*P. c. amoenissima*), Hutton's Vireo (*V. h. stephensi*), Bell's Vireo (*V. b. medius*), Hartlaub's Warbler, Summer Tanager (*P. r. cooperi*), Pine Siskin, Rufous-capped Atlaptes, and Black-throated Sparrow (*A. b. grisea*).

The Chihuahua-Zacatecas Biotic Province.--This province in Coahuila covers the arid, interior, western desert area; it consists of rolling plains with mountains that rise islandlike above the general surface. Some of the mountains, such as in the Sierra del Carmen and the Sierra del Pino, are more than 9000 feet high. The major part of this biotic area lies within the Lower Sonoran Life-zone. Areas of the Transition and Canadian life-zones are present on some of the higher mountains; their discontinuity results in a discontinuous distribution of the conifer-dependent avifauna.

The large desert restricts the movement of birds considerably. Major results of this include isolation of certain populations and absence of others in the boreal islands. For example, Miller (1955a:157) noted that the "dispersal of conifer-belt birds to and from the Sierra del Carmen, although not as difficult as to well separated islands [such as off the coast of Baja California], is nevertheless a formidable matter to accomplish across the great deserts of Texas, Chihuahua, and Coahuila." Miller (*loc. cit.*) noted also that the avifauna of the Sierra del Carmen, due to its insularity, is unbalanced and stated that "as a consequence of unbalance, species that are present show ecologic extension and unusual numerical relations." At least in this type of environment, an extension or expansion of the ecologic habits of the related types takes place when some species are absent.

This isolation influences local variation among some of the birds found in Coahuila. Niches elsewhere usually occupied by certain species, absent here, are occupied by other species. These other species thus enjoy an ecologic freedom and can expand their niches in the absence of related types of similar ecologic scope. For example, Miller (1955a:158-159) reported that Hairy Woodpeckers occurred only casually in the Sierra del Carmen and that the Ladder-backed Woodpecker has spread out and seems to occupy the niche or niches usually characteristic of the Hairy Woodpecker. Changes usually thought of as of subspecific character seem to be taking place between the Ladder-backed Woodpeckers of the Sierra del Carmen and of other areas, possibly because the Ladder-backed Woodpecker in the Sierra del Carmen is extending its ecologic sphere more than in areas where the Hairy Woodpecker exists. Restriction in dispersal due to geographic isolation has probably hindered gene flow, thus allowing rapid local adaptation, recognizable in variation at the infraspecific level. Miller (*loc. cit.*) listed other birds that have expanded their ecologic scope; his work should be referred to for further details.

The following birds are associated with this province: Black Vulture, Scaled Quail (*C. s. pallida*), Turkey, Elf Owl, Green Kingfisher, Hairy Woodpecker (*D. v. icastus*), Ladder-backed Woodpecker (*D. s. cactophilus*), Wied's Crested Flycatcher, Buff-breasted Flycatcher, Vermilion Flycatcher (*P. r. flammeus*), Black-crested Titmouse (*P. a. dysleptus*), Cactus Wren (*C. b. couesi*), Curve-billed Thrasher (*T. c. celsum*), Blue-gray Gnatcatcher (*P. c. amoenissima*), Hutton's Vireo (*V. h. carolinae*), Summer Tanager (*P. r. cooperi*), and Black-throated Sparrow (*A. b. opuntia*). Several kinds of birds, such as the Band-tailed Pigeon, occur in the "pine islands" in this province rather than on the desert floor.

There remain several kinds of birds that are not especially associated with any one or two of the above-named provinces. These birds are widely distributed and vary geographically without corresponding to the Biotic Provinces. Examples of these species are: Black Phoebe (*S. n. semiatra* in northern Coahuila; *S. n. nigricans* in southern Coahuila), Violet-green Swallow (*T. t. lepida* in northwestern Coahuila; *T. t. thalassina* in southeastern Coahuila), Black-eared Bushtit (*P. m. lloydi* in northern Coahuila; *P. m. iulus* in southeastern Coahuila), White-breasted Nuthatch (*S. c. nelsoni* in northern Coahuila; *S. c. mexicana* in southern Coahuila), Brown-throated Wren (*T. b. cahooni* in northern Coahuila; *T. b. compositus* in southern Coahuila), Crissal Thrasher (*T. d. dorsale* in northern Coahuila; *T. d. dumosum* in southern Coahuila), and Rufous-crowned Sparrow (*A. r. tenuirostris* in northern Coahuila; *A. r. boucardi* in southern Coahuila).

Some representatives of the avifauna of the central and southern sections of the Central Plateau reach southwestern Coahuila. The subspecies *squamata* of the Scaled Quail and *eurhyncha* of the Blue Grosbeak are examples. Each in Coahuila seems to be at the northern limit of its range.

In summary, there are three associations of vegetation in Coahuila and each has characteristic birds. Gross climate and topography, through their influence on vegetation, are the prime factors in the distribution and kinds of birds in the State. Some birds of central and southern Mexico reach southeastern and southwestern Coahuila. Representatives of the Gulf Coastal Plain in Tamaulipas and Nuevo Leon as well as migrants of the eastern flyway occur in northeastern Coahuila. Most of the species that occur in Coahuila seem to be associated with western North America. The aridity of western Coahuila restricts, to a large extent, the diversity of the breeding populations of its avifauna. Xeric conditions surrounding some of the higher mountains are barriers to movement of some species.

ORIGIN OF BREEDING BIRDS OF COAHUILA

Probably beginning in the late Pliocene and ending in the Ice Age (Griscom, 1950:379) the refrigeration of climate in the Northern Hemisphere initiated a period of southward withdrawal of birds from the northern part of North America. Some members of the avifauna of Coahuila probably reached the State in this time. When the continental deserts were formed, or reformed, many tropical and subtropical Middle American species were forced to leave Coahuila. Species associated with arid conditions found their way there. Many representatives of the Old World element also seem to have found their way to the State during the refrigeration of climate in the Northern Hemisphere. The separation of North and South America in the greater part of the Tertiary (Mayr,

1946:9) that deterred mammals from intercontinental colonization seemingly did not hinder birds. Some South American species moved northward into Mexico, all the way north to Coahuila.

The avifauna of Coahuila today is a mixture of the several mentioned elements. Of the breeding populations, 43 per cent breed in the western rather than the eastern United States, 6 per cent breed in the eastern rather than the western United States, 30 per cent breed in both the eastern and western United States, 20 per cent are restricted to the Republic of Mexico, and the southern parts of Arizona, New Mexico, and Texas, and 1 per cent (Aztec Thrush and Rufous-capped Atlapetes) is endemic to the Republic of Mexico.

It is instructive to consider also the origin of avifaunal elements at the level of Family. According to Mayr (1946:11) most North American families and subfamilies clearly originated in the Old World, in South America, or from a North American element that developed in the partial isolation of North America in the Tertiary. Three other elements, the Panboreal, the Pan-American, and the Pantropical are represented by some North American families and subfamilies. Because of the obscurity of the place of origin of certain groups, an additional unanalyzed element must be recognized.

The Caprimulgidae and Picidae probably originated in North America (Mayr, 1946:26). Although the Psittacidae are Pantropical in distribution, indications are that they probably originated in the Old World (Mayr, 1946:17). The Phasianidae, Turdidae (*Myadestes-Hylocichla* group), and Sylviidae (Polioptilinae) seem to have originated in the Old World (Mayr, 1946:27). However, Mayr considered these groups to have had a secondary center of proliferation in North America, and I thus consider these groups to have a North American origin. Mayr (1946:27) considered the Trochilidae, Tyrannidae, and Icteridae Pan-American in distribution; however, he suggested that they probably originated in South America, and I here treat them as South American in origin. No representatives of the Pan-American element that probably originated in North America have been recorded from Coahuila nor have members of the Panboreal element (Mayr, 1946:11) been recorded in the State. According to my analysis, representatives of families of birds known to breed in Coahuila and those that probably breed there thus seem to have been derived historically from the following sources:

Old World

24.7%

South America

24.0%

North America

37.0%

Unanalyzed

14.3%

Mayr (1946:28-29) gave examples of analysis by geographic origin of the breeding species of several districts of North America. For instance, at Yakutat Bay in southeastern Alaska the South American element of breeding passerine species was 3 per cent, the North American element 39 per cent, and the Old World element 58 per cent whereas at Sonora, Mexico, the South American element of breeding passerine species was 27 per cent, the North American element 52 per cent, and the Old World element 21 per cent. The breeding avifauna of Coahuila is thus in line with Mayr's analysis, resembling that of Sonora to a considerable degree at the taxonomic level of Family.

ACCOUNTS OF SPECIES

***Podiceps caspicus* (Hablizl).--On March 31, 1952, Olmstead saw "many Eared Grebes" on a pond 10 mi. E Hacienda La Mariposa. This is the first record of the Eared Grebe in Coahuila.

[***Pelecanus erythrorhynchos*** Gmelin.--The White Pelican is uncommon, if not rare; Friedmann, Griscom, and Moore (1950:21) list it.]

Anhinga anhinga (Linnaeus).--On March 31, 1952, Olmstead noted an Anhinga perched on a submerged fence post in a lake 10 mi. E Hacienda La Mariposa. This is the first record of the Anhinga in Coahuila.

*****Ardea herodias*** Linnaeus.--Two subspecies of the Great Blue Heron, *treganzai* and *wardi*, have been recorded from Coahuila. Friedmann, Griscom, and Moore (1950:27) listed *A. h. treganzai* from the State; presumably this subspecies occurs widely in low density. They (*loc. cit.*) remarked also that a record of *A. h. wardi* from Coahuila "cannot be allocated subspecifically."

Dickerman saw two Great Blue Herons in a marshy area at San Marcos (=20 mi. S Cuatro Cienegas) on May 4, 1954. Van Tyne and Sutton (1937:12) noted the Great Blue Heron "near Boquillas [Texas], along the Rio Grande, on May 10 and 15...."

*****Butorides virescens*** (Linnaeus).--Olmstead saw a Green Heron at Boquillas, 700 feet, on March 10, 1952. Findley reported seeing Green Herons 2 mi. W Jimenez, 850 feet, on June 19, 1952, and 2 mi. S and 3 mi. E San Juan de Sabinas on June 22, 1952.

*****Casmerodius albus egretta*** (Gmelin).--The Common Egret is an uncommon migrant in Coahuila. Friedmann, Griscom, and Moore (1950:30) recorded *C. a. egretta* from the "extreme northern part" of Coahuila. Van Tyne and Sutton (1937:12) saw the Common Egret "along the Rio Grande on the Graham ranch just west of Boquillas," Texas, on May 16, which might well be the locality to which Friedmann, Griscom, and Moore referred. Olmstead saw a Common Egret at Don Martin on March 22, 1952.

*****Nycticorax nycticorax hoactli*** (Gmelin).--This subspecies of the Black-crowned Night Heron was listed from the "extreme north" section of Coahuila by Friedmann, Griscom, and Moore (1950:32). Van Tyne and Sutton (1937:14) saw three Black-crowned Night Herons along the Rio Grande about two miles west of Boquillas, Texas, on May 16. This record probably represents the locality to which Friedmann, Griscom, and Moore (*op. cit.*) referred.

Nyctanassa violacea violacea (Linnaeus).--A Yellow-crowned Night Heron in immediate post-juvenile plumage, No. 36413, was obtained on September 7, 1958, 16 km. south of Cuatro Cienegas, by W. L. Minckley. According to him the bird was accompanied by "several" other herons seemingly of the same species and condition of plumage. The species seems not to have been recorded previously from Coahuila [Eds.].

Botaurus lentiginosus (Rackett).--Friedmann, Griscom, and Moore (1950:34) listed the American Bittern from the "extreme northern part" of Coahuila. Van Tyne and Sutton (1937:14) saw two representatives of this bittern "along the Rio Grande not far from Hot Springs," Texas, on May 15. I suspect that this is the locality to which Friedmann, Griscom, and Moore (*loc. cit.*) referred.

[***Branta canadensis leucopareia*** (Brandt).--Friedmann, Griscom, and Moore (1950:38) listed *B. c. leucopareia* from "northern Coahuila."]

*****Dendrocygna autumnalis*** (Linnaeus).--Evenden (1952:112) reported a Black-bellied Tree Duck standing beside a reservoir in southern Coahuila along the railroad between Saltillo, Coahuila, and Avalos, Zacatecas.

*****Anas platyrhynchos*** Linnaeus.--On March 30, 1952, Olmstead recorded a Mallard from 10 mi. E Hacienda La Mariposa, 2000 feet.

Anas strepera Linnaeus.--*Specimen examined:* one, 31016, from 10 mi. E Hacienda La Mariposa, 2000 ft., March 30, 1952.

The Gadwell is not an uncommon spring migrant; Olmstead saw it 10 mi. E Hacienda La Mariposa on March 30, 1952, and Baker observed it 8 mi. N and 4 mi. W Muzquiz on March 30, 1952. Friedmann, Griscom, and Moore (1950:42) recorded the Gadwell from Coahuila.

Anas acuta Linnaeus.--Miller (1955a:161) took a Pintail on September 10 in the Sierra del Carmen.

Anas carolinensis Gmelin.--The Green-winged Teal has been recorded from northern Coahuila. Van Tyne and Sutton (1937:15) recorded two mated pairs along the Rio Grande at Lajitas, Texas, on May 10. Miller (1955a:161) remarked that a male of the year was taken in the Sierra del Carmen on September 4.

Anas discors discors Linnaeus.--*Specimens examined:* total 2: sex ? 31646 and sex ? 31647 from .5 mi. S Las Margaritas, 2800 ft., September 28, 1953.

The Blue-winged Teal is a fairly common spring and fall migrant in Coahuila. Van Tyne and Sutton (1937:15) noted the Blue-winged Teal at several different localities along the Rio Grande: "on May 8, four males and several females resting on a mud bar along the Rio Grande near Hot Springs [Texas]; ... on May 7, three pairs in a flock, along the Rio Grande, Castalon [Texas]; ... and on May 20, three pairs, along the Rio Grande, San Vicente [Texas]." Miller (1955a:161) reported that Marsh took a male of the year in the Sierra del Carmen on September 10. Dickerman observed Blue-winged Teal 8 mi. E and 2 mi. S Americanos on May 18, 1954. Olmstead listed Blue-winged Teal from 10 mi. E Hacienda La Mariposa on March 30, 1952. Nos. 31646-31647, which are probably females, represent the subspecies *discors* because the light edgings of their crowns are definitely present; the areas of their backs are brownish, not more intensively black, and their underparts are brownish, less blackish.

*****Anas cyanoptera septentrionalium*** Snyder and Lumsden.--Van Tyne and Sutton (1937:15) listed several localities along the Rio Grande in Brewster County, Texas, where Cinnamon Teal were seen. I suspect that Friedmann, Griscom, and Moore (1950:41) referred to those localities. Dickerman saw four pairs of Cinnamon Teal 14 mi. E and 16 mi. N Ocampo on May 9, 1954, and also saw Cinnamon Teal 8 mi. E and 2 mi. S Americanos on May 18, 1954.

Mareca americana (Gmelin).--The American Widgeon is a fairly common spring migrant in Coahuila. Olmstead observed this duck 10 mi. E Hacienda La Mariposa on March 30, 1952. Dickerman saw five to seven American Widgeons 8 mi. E and 2 mi. S Americanos on May 18, 1954.

Spatula clypeata (Linnaeus).--The Shoveler is a spring and probably fall migrant in Coahuila, and has been observed at several localities. Van Tyne and Sutton (1937:16) saw two pairs along the Rio Grande at Castalon, Texas, On May 7 and saw "a fair-sized flock along the Rio Grande on the Johnson ranch [in Texas] on May 13 and 14." Dickerman saw 12 pairs of Shovelers on two ponds 14 mi. E and 16 mi. N Ocampo on May 9, 1954, and 10 more 8 mi. E and 2 mi. S Americanos on May 18.

Aythya affinis (Eyton).--Olmstead observed Lesser Scaup 10 mi. E Hacienda La Mariposa on March 30, 1952.

[***Bucephala albeola*** (Linnaeus).--Friedmann, Griscom, and Moore (1950:44) listed the Bufflehead from the State.]

****Cathartes aura aura*** (Linnaeus).--*Specimen examined:* one, 31017 (skeleton only), from 4 mi. W Hacienda La Mariposa, 2300 ft., March 26, 1952.

Miller (1955a:161) took a female Turkey Vulture, which was in breeding condition, in the Sierra del Carmen on April 17 and stated that "until more statistics are available on breeding birds of northern Coahuila, they must be considered *C. a. aura*...." Amadon and Phillips (1947:577) took a Turkey Vulture at Las Delicias which represented *C. a. aura*. Burleigh and Lowery (1942:188) stated that this species was not uncommon, and was noted each day soaring overhead both in the valleys and over the tops of the ridges of southeastern Coahuila.

Friedmann, Griscom, and Moore (1950:47) listed *C. a. teter* from Coahuila. Miller (1955a:161) remarked that the subspecies *aura* and *teter* might intergrade in the Sierra del Carmen. At the present time it is possible to say only that *teter* is present in Coahuila in migrant and wintering populations, but the extent to which *teter* remains in northeastern Mexico is undetermined. However, all indications point to this area as being the region where *aura* and *teter* intergrade.

*******Coragyps atratus* (Bechstein).--The Black Vulture is locally common throughout most of eastern Coahuila but is uncommon in the western part of the State. Sutton and Burleigh (1939a:25) noted the Black Vulture "regularly east of Saltillo in low country," but did not see Black Vultures at San Pedro or elsewhere in southwestern Coahuila. Burleigh and Lowery (1942:188) stated that "the Black Vulture apparently avoids to a large extent the higher altitudes, and only rarely was it observed at all, even about Saltillo." Olmstead saw Black Vultures 8 mi. N and 4 mi. W Muzquiz, 1800 feet, on March 31, 1952, and Dickerman observed a flock at La Gacha (=Rancho La Coucha), 1600 feet, on December 2, 1953.

****_Accipiter gentilis_** (Linnaeus).--On July 6, 1955, Hardy saw a Goshawk 13 mi. E San Antonio de las Alazanas; this is the first record of occurrence of this species from northeastern Mexico.

***_Accipiter striatus velox_** (Wilson).--*Specimen examined:* one, 31018, from along the Rio Grande (=Boquillas), 700 ft., March 10, 1952, measurements: wing, 207 mm.; tail, 171 mm.; tarsus, 53 mm.; culmen, 12 mm.

Our specimen of the Sharp-shinned Hawk is referred to *velox* on the basis of the reddish, maculated breast, sides, and thighs. The collector's field notes recorded the iris as blood-red. Marsh and Stevenson (1938:286) thought that this subspecies was resident in the pine and Douglas-fir forest of upper Vivoras Canyon of the Sierra del Carmen at 8500 feet, where Marsh observed a family group including three immature birds. Friedmann (1950:196) indicated that the immature male obtained by Marsh and Stevenson is *A. s. suttoni*; Miller (1955a:161), nevertheless, remarked that this male has well barred feathers and thus is *velox*. Miller (*loc. cit.*) obtained also an adult male of *A. s. velox* in the Sierra del Carmen at 7000 feet on April 18.

****Accipiter striatus suttoni** van Rossem.--*Specimen examined:* one, 32626, from 13 mi. E San Antonio, 9950 ft., July 6, 1955, measurements: wing, 186 mm.; tail, 144 mm.; tarsus, 49 mm.; culmen, 11 mm.; weight, 103 gms.

The recording of *A. s. suttoni* in Coahuila by Friedmann, Griscom, and Moore (1950:52), seems to have been based on their knowledge of the specimen earlier mentioned by Friedmann (1950:196) and later identified by Miller (1955a:161) as *A. s. velox*. Therefore the KU specimen seems to be the first record of *A. s. suttoni* in Coahuila. The size of its testes (right, 2.5x4 mm.; left, 3x4 mm.) does not indicate breeding; however, the time of the year in which it was obtained suggests that it may have been a resident.

***Accipiter cooperii** (Bonaparte).--Miller (1955a:161) found Cooper's Hawk breeding in the Sierra del Carmen on April 26.

Buteo jamaicensis borealis (Gmelin).--The Red-tailed Hawk is common in Coahuila. Burleigh and Lowery (1942:188) noted the Red-tailed Hawk on the higher ridges above an elevation of 6000 feet in southeastern Coahuila. On April 17, Burleigh and Lowery (*loc. cit.*) saw two Red-tailed Hawks "in the open valley south of Diamante Pass" and on April 20, "just outside of Saltillo," these workers obtained an immature male that was referred to *B. j. borealis*.

****Buteo jamaicensis fuertesi** Sutton and Van Tyne.--Miller (1955a:161) took a male Red-tailed Hawk, on April 14 at 7000 feet in the Sierra del Carmen, that was referred to as *B. j. fuertesi*. To my knowledge, there are no other records of this subspecies from Coahuila, but this must be the resident form over the bulk of western Coahuila.

There are several sight records of the Red-tailed Hawk. Olmstead saw one 16 mi. S Boquillas, 1600 feet, on March 6, 1952; Dickerman saw a Red-tailed Hawk 16 mi. E and 18 mi. N Ocampo on May 7, 1954, one 20 mi. S Ocampo on April 4, 1954; and an immature at Saltillo on January 17, 1954.

Buteo platypterus platypterus (Vieillot).--*Specimen examined:* one, 32628, from 13 mi. E San Antonio de las Alazanas, 9950 ft., July 6, 1955.

The Broad-winged Hawk is rare in Coahuila. No. 32628, if a migrant, was retarded from moving northward by the loss of its right foot and distal one-third of its tarsus. Packard (1957:371) reported this specimen as the first record of the species in Coahuila.

***_Buteo swainsoni_ Bonaparte.**--*Specimens examined:* total 2: 32022 from 2 mi. W Jimenez, 850 ft., June 20, 1952; and 29555 from Iglesias (=15 mi. SW Sabinas), 1000 ft., August 22, 1949.

Swainson's Hawk is not common in Coahuila. The size of the testes (6x4 mm.) of No. 32022, the adult plumage, and the date (June 20) on which it was obtained suggest that it was a breeding bird. Friedmann, Griscom, and Moore (1950:55) reported that this species breeds as far east as Durango and Chihuahua. Findley saw a Swainson's Hawk 2 mi. S and 3 mi. E San Juan de Sabinas on June 22, 1952.

****_Buteo albonotatus_ Kaup.**--The Zone-tailed Hawk is uncommon in Coahuila. Sutton and Burleigh (1939a:26) noted the species "a few miles west of Saltillo ... on January 30."

Buteo regalis (Gray).--The Ferruginous Hawk is uncommon in Coahuila. Sutton and Burleigh (1939a:26) noted a single bird "not far from Parras, on January 30."

Buteo nitidus (Latham).--Evenden (1952:112) saw a Gray Hawk one mile northeast of Saltillo, at Ramos Arizpe on March 4. Although there are no other records of the Gray Hawk from Coahuila, its occurrence in the State would be expected because this species has been recorded from Nuevo Leon and Tamaulipas in northeastern Mexico (Friedmann, Griscom, and Moore, 1950:57).

****_Parabuteo unicinctus harrisi_** (Audubon).--Harris' Hawk is fairly common in southern Coahuila. Sutton and Burleigh (1939a:26) stated that Harris' Hawk "was one of the few birds noted repeatedly in ... southern Coahuila...."

Circus cyaneus (Linnaeus).--The Marsh Hawk is a common migrant and winter visitant in Coahuila. Miller (1955a:161) remarked that the Marsh Hawk was "seen in northward migratory flight across the desert east of the Sierra del Carmen on March 31 and on April 11 along the west face of Loomis Peak at 8800 feet." Olmstead saw a female Marsh Hawk 1 mi. W San Buenaventura on April 2, 1952. Burleigh and Lowery (1942:188) noted the Marsh Hawk "about Saltillo" and "above the summit of Diamante Pass at about 8,000 feet" on April 14. Sutton and Burleigh (1939a:26) noted wintering Marsh Hawks "near San Pedro."

Pandion haliaetus (Linnaeus).--Miller (1955a:161) reported seeing an Osprey on April 9 in Corte Madera Canyon, "apparently in migration;" this is the only record from Coahuila.

****_Caracara cheriway_** (Jacquin).--From the few records in the literature, I judge that the Caracara is uncommon in Coahuila. Evenden (1952:113) saw three Caracaras "south of Saltillo" on March 5. Baker saw a Caracara in the Sierra del Pino (=6 mi. NW Tanque Alvarez), 3400 feet, on July 6, 1953. No specimens of the Caracara have been taken from Coahuila.

Falco mexicanus Schlegel.--*Specimen examined:* one, 31596, from Saltillo, January 10, 1954.

The Prairie Falcon is an uncommon winter visitant in Coahuila. Friedmann, Griscom, and Moore (1950:65) indicated that *Falco mexicanus* winters south to Sonora, Oaxaca, Chihuahua, Durango, Zacatecas, Auguascalientes, Hidalgo, Nuevo Leon, and Tamaulipas.

No. 31596 is the first recorded specimen of the Prairie Falcon from Coahuila. The bird was heavily parasitized by worms in the mesenteries and seems to be an adult. Although its nuchal collar, as in immatures, is washed with pale cinnamon-buff, its thighs are not heavily marked with dark brown spots. The superciliary lines have blackish rather than brownish streaks, and the scapulars do not have four or five dark bars (Friedmann, 1950:624).

Falco columbarius bendirei Swann.--*Specimen examined:* one, 31634, from Don Martin Dam (=Rio Salado), November 27, 1953, measurements: wing, 191 mm.; tail, 111 mm.; tarsus, 37 mm.; culmen, 12 mm.; testes, 3x1 mm.

The Pigeon Hawk seems to be uncommon in Coahuila. No. 31634, the first record of this species in Coahuila, has pale gray interspaces on the rectrices of its tail that are definitely wider than the three black bands, indicating affinity with *bendirei* (Friedmann, 1950:702). Our bird was obtained near the base of the Don Martin Dam of the Rio Salado, and was observed hunting dragonflies over the water.

Falco sparverius sparverius Linnaeus.--*Specimen examined:* one, 31648, from the north foot of Sierra Guadalupe (=10 mi. S, 5 mi. W General Cepeda), 6400 ft., April 17, 1953, weight, 104 gms.

The Sparrow Hawk is locally common in Coahuila. Miller (1955a:162) noted the species occasionally in the lower canyon areas of the Sierra del Carmen at 5000 feet from April 20 to 28. Dickerman saw two Sparrow Hawks in the Sierra del Pino on May 12, 1954. Sutton and Burleigh (1939a:27) took a male at La Rosa on January 30 that was typical for *F. s. sparverius*. Burleigh and Lowery (1942:188) noted Sparrow Hawks almost daily about Saltillo and infrequently in the open valley south of Diamante Pass; they took a female at Saltillo on April 22 that was assigned to *F. s. sparverius*.

****Colinus virginianus texanus*** (Lawrence).--*Specimens examined:* total 9: 29408 and 29409 from 1 mi. S, 9 mi. W Villa Acuna, April 2, 1950, weights, 169.5 and 174.7 gms., 29410 and 29411-29413 from 3 mi. W, 1 mi. S San Geronimo, April 9, 1950, weights, 152.6, 158.6, 158.2, 159.0, and 152.8 gms.; and 32032-32034 from 9 mi. S, 11 mi. E Sabinas, 1050 ft., June 13 and 14, 1952.

The Bobwhite is locally common in northeastern Coahuila. Aldrich and Duvall (1955:18) indicated that *C. v. texanus* has been recorded from two separate localities in northeastern Coahuila and from several localities in Texas along the Rio Grande. Findley saw Bobwhites 2 mi. S and 3 mi. E San Juan de Sabinas on June 22, 1952. The specimens of Bobwhite from 3 mi. W and 1 mi. S San Geronimo seem to extend the known range of *texanus* nearly 100 miles west. The sizes of the testes (11, 16, 15, 15 mm.) of Nos. 29408, 32032-32034, respectively, indicates breeding by the Bobwhite in Coahuila.

***Callipepla squamata pallida** Brewster.--*Specimens examined:* total 16: 29414 from 1 mi. S, 13 mi. W Villa Acuna, April 4, 1950; 29415-29417 from 11 mi. W Hacienda San Miguel, 2200 ft., April 4, 1950, weights, 205.3, 198.6, and 182.7 gms.; 31019-31021 from 10 mi. S, 5 mi. E Boquillas, 1500 ft., March 2 and 4, 1952, weights, 184.4, 180 and 154.2 gms.; 34454 from 2 mi. SSE Castillon, 4050 ft., June 29, 1953, weight, 169 gms.; 29418 and 29419 from 8 mi. N, 2 mi. E La Babia, April 8, 1950; 32023-32024, 32026, 32025, and 32027 from 5 mi. N, 19 mi. W Cuatro Cienegas, 3250 ft., July 6, 1952; and [** Male] 32640 from 2 mi. W Paila, July 3, 1955.

The Scaled Quail is common in Coahuila. The subspecies *pallida* occurs in northwestern Coahuila south to Sierra de los Alamitos. Intergrades of *pallida, castanogastris,* and *squamata* are present in the central part of the State. No. 32640, obtained 2 mi. W Paila, has some resemblance to *squamata*. Five specimens of *pallida* from the central part of Coahuila (5 mi. N and 19 mi. W Cuatro Cienegas), show little or no approach toward *squamata*. Miller (1955a:162) stated that two of the Scaled Quail collected in the Sierra del Carmen show no approach to the race *castanogastris* of eastern Coahuila nor to *C. s. squamata* of southern Coahuila. From the specimens that I have examined, I judge that the range of *pallida* extends as far south as the Sierra de los Alamitos rather than only to the northwestern part of Coahuila as reported by Aldrich and Duvall (1955:17). In northeastern Coahuila *pallida* seems to intergrade with *castanogastris*; No. 29414 has an indistinct rusty chestnut patch on its abdomen, thus resembling *castanogastris*.

The sizes of the testes (9-12 mm.) and of the largest ova (14 mm. in diameter and an egg 23 mm. long) of birds labeled with reference to Cuatro Cienegas indicate breeding activity.

****Callipepla squamata castanogastris** Brewster.--*Specimen examined:* one, 32028, from 9 mi. S, 11 mi. E Sabinas, June 14, 1952.

Typical representatives of *C. s. castanogastris* in Coahuila occur only in the extreme northeastern section of the State, and most specimens of the Scaled Quail from northeastern Coahuila are intergrades between *pallida* and *castanogastris*.

No. 32028 is identified as *C. s. castanogastris* because there is a distinct rusty chestnut patch on its abdomen. This patch, nevertheless, is not so large as in a more nearly typical male *C. s. castanogastris* from 15 mi. NNW Anahuac, Nuevo Leon.

****Callipepla squamata squamata** (Vigors).--*Specimens examined:* total 2: 30231 and 30232 from 10 mi. NW San Lorenzo, 4200 ft., February 3, 1951.

The subspecies *squamata* occurs in southern Coahuila. Amadon and Phillips (1947:577) took a Scaled Quail at Las Delicias on August 18 that "was only two-thirds grown, though well able to fly" and obtained an adult 19 mi. W Saltillo that was typical *squamata*. Burleigh and Lowery (1942:188) stated that *C. s. squamata* was one of the characteristic birds of the open desert country of southeastern Coahuila. Scaled Quail were seen by Burleigh and Lowery (*loc. cit.*) "about Saltillo and in the open valley south of Diamante Pass."

The breast and upper back of both specimens from 10 mi. NW San Lorenzo, are plumbeous-gray rather than pale dull gray. Also the lower back, rump, abdomen, forehead, and crown more closely resemble the subspecies *squamata* rather than *C. s. pallida*. However, the upper backs of both specimens are not so plumbeous-gray as on a male (32030) and a female (32031) of the subspecies *squamata* from 1 mi. N Chorro, Durango, 6450 ft., July 11, 1952. The two birds from Durango appear to be slightly darker than the Coahuilan specimens that approach the subspecies *pallida*.

Burleigh and Lowery (1942:188-189) stated that one of their specimens of *C. s. squamata* obtained at Saltillo seems to be "very close to *castanogastris*, suggesting that southeastern Coahuila is in the region of intergradation between the two races." Aldrich and Duvall (1955:17) indicated that *squamata* and *castanogastris* intergrade near Sabinas. The two specimens that I have examined show no sign of approach toward *castanogastris*. More specimens of Scaled Quail from Coahuila are needed to permit accurate definition of the distribution of the subspecies.

***Cyrtonyx montezumae mearnsi** Nelson.--The Harlequin Quail is locally common in Coahuila; *C. m. mearnsi* is present in northwestern

Coahuila (Aldrich and Duvall, 1955:20). Miller (1955a:162) stated that an area in the head of Corte Madera Canyon of the Sierra del Carmen at 7500 feet was the principal location for *C. m. mearnsi*. He further suggested that the Harlequin Quail breeds in the Sierra del Carmen and remarked that Marsh took a male on September 7 at Jardin del Sur. He added that the occurrence of *C. m. montezumae* in northern Coahuila as reported by Friedmann, Griscom, and Moore (1950:79) is "possibly an error or was based on individual dark variants...." Baker noted the Harlequin Quail (unidentified to subspecies) 9 mi. W and 1 mi. S Villa Acuna, 1120 feet, on April 4, 1950.

****Cyrtonyx montezumae montezumae*** (Vigors).--This subspecies of the Harlequin Quail has been recorded from southeastern Coahuila. Ridgway and Friedmann (1946:396) listed *C. m. montezumae* from Saltillo. Baker saw a pair of Harlequin Quail (unidentified to subspecies) at San Antonio de las Alazanas on March 25, 1950. More collecting is necessary for an understanding of the distribution and intergradation of these subspecies in Coahuila.

****Meleagris gallopavo intermedia*** Sennett.--*Specimens examined:* total 4: 31022 from Fortin (=33 mi. N, 8 mi. W San Geronimo), 3300 ft., March 28, 1952, weight, 9 lbs.; 29420 from 3 mi. W, 1 mi. S San Geronimo, April 9, 1950, weight, 7 lbs.; and 29421 and 29422 from 3 mi. W, 2 mi. S San Geronimo, April 9, 1950, weights, 11.5 and 8.5 lbs.

The Turkey in Coahuila is restricted to the northern section of the State. Miller (1955a:162) remarked that the population of Turkeys in the Sierra del Carmen was sparse and did not range above 7500 feet into the highest pine-oak and Douglas fir areas. Baker saw Turkeys 4 mi. W Hacienda La Mariposa, 2300 feet, on March 23, 1952. William Schaldach, Jr., noted 30 Turkeys "just west of Rancho San Geronimo" on April 9, 1950. Aldrich and Duvall (1955:22) indicated several localities in northern Coahuila where the Turkey occurs or occurred.

Grus canadensis (Linnaeus).--Sutton and Burleigh (1939a:28) remarked that a flock of Sandhill Cranes was heard "near Mayran ... on January 30."

Porzana carolina (Linnaeus).--The Sora is an uncommon spring and fall migrant in Coahuila. Amadon and Phillips (1947:577) obtained an adult male Sora at Las Delicias on August 15.

Fulica americana Gmelin.--The American Coot probably occurs in the State as a spring and fall migrant. Dickerman saw two coots 8 mi. E and 2 mi. S Americanos on May 18, 1954.

**Charadrius vociferus vociferus* Linnaeus.--*Specimen examined:* one, 31023, from the Rio Grande (=17 mi. S Dryden, Terrell Co., Texas, in Coahuila), 600 ft., March 19, 1952, weight, 82.7 gms.

In Coahuila the Killdeer seems to be uncommon. Van Tyne and Sutton (1937:28) remarked that it nested in Brewster County, Texas. The species probably nests in northern Coahuila as well; the lengths of the testes (left, 9 mm.; right, 4 mm.) of No. 31023 support this view.

Eupoda montana (Townsend).--Pitelka (1948:118) recorded one female Mountain Plover from Hipolito on February 23. Van Tyne and Sutton (1937:28) reported that the Mountain Plover nested in Brewster County, Texas. Possibly *Eupoda montana* nests in northern Coahuila as well.

Numenius americanus parvus Bishop.--*Specimens examined:* total 3: 31434 and 35406 (skeleton only) from 8 mi. E, 2 mi. S Americanos, May 19, 1954, measurements: wing, 250, 258 mm.; tail, 96, 100 mm.; tarsus, 74, 81.5 mm.; culmen, 11.2, 16.5 mm.; gonad, 6x2 mm.; and 35400 (skeleton only) from 7 mi. W San Antonio de las Alazanas, January 11, 1954.

The Long-billed Curlew is not common in Coahuila. Dickerman obtained both the specimens from 8 mi. E and 2 mi. S Americanos out of a flock of 35. Sutton and Burleigh (1939a:28) noted the Long-billed Curlew "near San Pedro ... [on] January 29."

Actitis macularia (Linnaeus).--The Spotted Sandpiper has been obtained from two localities in Coahuila. Miller (1955a:162) stated that an immature in fall migration was taken "at the tank in the western hills" of the Sierra del Carmen on September 4. Burleigh and Lowery (1942:189) found the Spotted Sandpiper "at the Chorro del Agua near Arteaga" on April 17 and 19.

Totanus melanoleucus (Gmelin).--*Specimen examined:* one, 31024, from the Rio Grande (=17 mi. S Dryden, Terrell Co., Texas, in Coahuila), 600 ft., March 18, 1952, weight, 224 gms.

The Greater Yellowlegs is an uncommon spring and probably fall migrant in Coahuila. No. 31024 is the first record of this species in Coahuila.

Totanus flavipes (Gmelin).--Miller (1955a:162) reported that Marsh took a Lesser Yellowlegs "at the tank in the western hills" of the Sierra del Carmen on September 4.

Erolia melanotos (Vieillot).--Miller (1955a:162) reported taking a Pectoral Sandpiper on September 4 "at the tank in the western hills" of the Sierra del Carmen.

[***Erolia minutilla*** (Vieillot).--Friedmann, Griscom, and Moore (1950:99) listed the Least Sandpiper from Coahuila.]

****Recurvirostra americana*** Gmelin.--*Specimens examined:* total 2: 31433 and 31432 from 8 mi. E, 2 mi. S Americanos, May 19, 1954.

Van Hoose (1955:302) reported a small breeding colony of American Avocets 8 mi. E and 2 mi. S Americanos "on a large grassy playa traversed by rows of creosote (*Larrea tridentata*)." No. 31432 was taken from a nest containing four partly-incubated eggs. Van Hoose (*loc. cit.*) also reported that four eggs in a second nest contained well-developed, downy young.

Steganopus tricolor Vieillot.--Wilson's Phalarope occurs in Coahuila as a spring and probably fall migrant. Van Tyne and Sutton (1937:31) saw the species several times along the Rio Grande. Friedmann, Griscom, and Moore (1950:102) listed *Steganopus tricolor* from the Rio Grande along the borders of Chihuahua and Coahuila on May 10-16.

[***Larus argentatus smithsonianus*** Coues.--Friedmann, Griscom, and Moore (1950:104) listed this subspecies of the Herring Gull from the State.]

[***Larus delawarensis*** Ord.--Friedmann, Griscom, and Moore (1950:103) stated that the Ring-billed Gull occurs in Coahuila.]

****Columba fasciata fasciata*** Say.--*Specimens examined:* total 3: 35401 (skeleton only) from 22 mi. S and 5 mi. W Ocampo, 7000 ft., April 6, 1954; and 32035 and 32036 from 4 mi. N, 21 mi. W Cuatro Cienegas, 6200 ft., July 2, 1952.

The Band-tailed Pigeon is locally common in Coahuila. Miller (1955a:162) remarked that the Band-tailed Pigeon "was unaccountably rare in 1953" in the Sierra del Carmen, and said that a specimen was taken on August 7 in Vivoras Canyon. Burleigh and Lowery (1942:189) wrote that "small flocks were seen each day ... on the summit of Diamante Pass, but only on one occasion was a pair noted." Dickerman saw 10 Band-tailed Pigeons 20 mi. S Ocampo, 6000 feet, on April 4, 1954. The enlarged testes (17 mm.) of No. 32035, and an egg (38 mm. long) in No. 32036 show that the species breeds in central Coahuila.

****Zenaida asiatica asiatica*** (Linnaeus).--*Specimen examined:* one, 31025 (skeleton only), from 10 mi. S and 5 mi. E Boquillas, 1500 ft., March 4, 1952.

Miller (1955a:163) took a male White-winged Dove on April 23 "at the mouth of Boquillas Canyon at about 4900 feet" and remarked that this bird might either be a straggler or a representative of a normal breeding population (Z. *a. asiatica*[**Z doesn't look italic, but should be?]). On

January 29 and 30, Sutton and Burleigh (1939a:29) saw White-winged Doves several times in southern Coahuila along the highway from Saltillo to San Pedro.

*****Zenaida asiatica mearnsi*** (Ridgway).--Amadon and Phillips (1947:577) obtained an adult White-winged Dove of this subspecies at Delicias on August 14.

Zenaidura macroura carolinensis (Linnaeus).--Friedmann, Griscom, and Moore (1950:117) listed this subspecies of the Mourning Dove as a spring and fall migrant on the "Central Plateau" (Coahuila is part of the "Central Plateau,"), "as proven by banding records not indicated in literature."

****Zenaidura macroura marginella*** (Woodhouse).--*Specimen examined:* one, 34455, from 1 mi. W Jaco, June 27, 1953.

Miller (1955a:163) reported *Z. m. marginella* as occurring commonly in the desert border and lower canyons at the base of the mountains of the Sierra del Carmen, "but it occasionally ranged up to 6000 feet in openings in the pine-oak belt." Burleigh and Lowery (1942:189) remarked that this dove was fairly plentiful in the open desert country "about Saltillo, and was ... noted in small numbers in Diamante Valley on April 17 and 19;" they also said that the Mourning Dove was not seen above an elevation of about 7500 feet. Sutton and Burleigh (1939a:29) observed Mourning Doves along the highway across southern Coahuila. Mourning Doves were seen by Findley 2 mi. W Jimenez, 850 feet, on June 19, 1952, and 2 mi. S and 11 mi. E Nava, 810 feet, on June 15, 1952. Dickerman saw one in the Sierra del Pino on May 12, 1954. Findley saw more than one 2 mi. S and 3 mi. E San Juan de Sabinas on June 22, 1952, as did Dickerman at San Marcos (=20 mi. S Cuatro Cienegas) on May 4, 1954.

*****Columbigallina passerina*** (Linnaeus).--The Ground Dove seems to be uncommon in Coahuila. Van Tyne and Sutton (1937:34) saw a single Ground Dove fly across the Rio Grande into Coahuila at Lajitas, Texas, on May 10. Findley saw one 2 mi. S and 3 mi. E San Juan de Sabinas on June 22, 1952.

*****Scardafella inca*** (Lesson).--The Inca Dove has been recorded from two localities in Coahuila. Hellmayr and Conover (1942:510) listed it from Sabinas. Burleigh and Lowery (1942:189) collected a male and female on April 16 and 19, respectively, "outside the city limits of Saltillo."

Leptotila verreauxi angelica Bangs and Penard.--*Specimens examined:* total 2: 31026-31027 (skeletons only) from 4 mi. W Hacienda La Mariposa, 2300 ft., March 25, 1952.

The White-fronted Dove seems to be uncommon in Coahuila. Hellmayr and Conover (1942:570) listed *L. v. angelica* from Sabinas.

****Rhynchopsitta pachyrhyncha terrisi** Moore.--*Specimens examined:* total 4: 31531-31532 from 13 mi. E San Antonio de las Alazanas, 9345 ft., April 10, 1954, weights, 391.5 and 467.5 gms.; 31533 from 13 mi. E San Antonio de las Alazanas, 10,000 ft., April 10, 1954, weight, 466 gms.; and sex ? 31534 from Mesa de las Tablas, June, 1951.

The Thick-billed Parrot occurs in the southeastern section of the State, where it is fairly common. Moore (1947:27-28) described this parrot as *Rhynchopsitta terrisi*: he thought it differed decidedly from *Rhynchopsitta pachyrhyncha*. However, Hardy and Dickerman (1955:305-306) decided that uniting the two forms as a single species better expresses their relationship.

Burleigh and Lowery (1942:189) reported seeing a small flock of Thick-billed Parrots on the summit of Diamante Pass. Dickerman, in his field notes, wrote that at a place 13 mi. E San Antonio de las Alazanas, 9345 feet, a large flock of about 300 birds was in a spruce-fir-pine-aspen association.

***Coccyzus americanus americanus** (Linnaeus).--*Specimens examined:* total 2: 32037-32038 from 12 mi. N, 12 mi. W Jimenez, 850 ft., June 19, 1952, measurements: wing, 141, 146 mm.; tail, 142, 149 mm.; tarsus, 27, 27 mm.; culmen, 25, 24 mm.

In Coahuila, the Yellow-billed Cuckoo seems to be uncommon. It occurs in the northeastern section of the State, in the Gulf Coastal Plain (Baker, 1956:128), and probably breeds there. One subspecies, *americanus,* has been recorded from Coahuila.

According to Ridgway (1916:13-17) the difference between *C. a. americanus* and *C. a. occidentalis* is size. His (*loc. cit.*) average measurements of males of *occidentalis* are: wing, 149.6 mm.; tail, 147.1 mm.; tarsus, 26.7 mm.; and culmen, 27.7 mm. whereas average measurements given by him of males of *americanus* are: wing, 143.6 mm.; tail, 140.7 mm.; tarsus, 25.2 mm.; and culmen, 26.4 mm. Van Tyne and Sutton (1937:35) question the value of maintaining the subspecies *occidentalis,* because individuals of *americanus* and *occidentalis* are almost impossible to tell apart. Friedmann, Griscom, and Moore (1950:132) stated that *americanus* occurs in eastern North America whereas *occidentalis* occurs in western North America. If the subspecies *occidentalis* exists, then Nos. 32037 and 32038 are, by size, *americanus* and No. 32038 is an intergrade between the two subspecies (or a large individual of *americanus*).

The Yellow-billed Cuckoo was seen also by Findley 2 mi. S and 3 mi. E San Juan de Sabinas on June 22, 1952, and by Dickerson at Torreon on July 2, 1955. The sizes of the testes of the birds from 12 mi. N and 12 mi. W Jimenez (9, 10 mm. long) and the date (June 19) on which they occurred there indicate that the birds possibly were breeding.

Coccyzus erythropthalmus (Wilson).--Miller (1955a:163) reported a migrant Black-billed Cuckoo taken in the maples and basswood near a water hole in the bottom of Boquillas Canyon in the Sierra del Carmen, 5200 feet, on April 22. Friedmann, Griscom, and Moore (1950:132) reported that this cuckoo is presumably a regular transient in Mexico, but generally overlooked.

****Geococcyx californianus*** (Lesson).--*Specimen examined:* one, 32049, from 8 mi. N, 2 mi. W Piedras Negras, June 18, 1952.

Miller (1955a:163) heard several Roadrunners calling at Boquillas Canyon in the Sierra del Carmen, where he obtained two females. Burleigh and Lowery (1942:190) stated that the species proved to be "unexpectedly scarce" and was noted but once by them on April 22 when a single bird was observed in "the open desert west of Saltillo." Sutton and Burleigh (1939a:30) noted that the Roadrunner was not common anywhere in southern Coahuila; they obtained one female at San Pedro on January 29. The size of the largest ovum (15 mm. in diameter) of No. 32040 indicates that this species breeds in Coahuila.

****Crotophaga sulcirostris sulcirostris*** Swainson.--*Specimen examined:* one, 32039, from 2 mi. S, 3 mi. E San Juan de Sabinas, June 22, 1952.

No. 32039, obtained by Harrison B. Tordoff in a cypress woods along the shore of a lagoon, provides the first record of the Groove-billed Ani in Coahuila. The size of its largest ovum (10 mm. in diameter) and the date indicate breeding by this species in Coahuila.

*****Tyto alba pratincola*** (Bonaparte).--The Barn Owl seems to be uncommon in Coahuila. To my knowledge, there are two records of the Barn Owl in Coahuila. Ridgway (1914:607) recorded this owl at the "head of Las Vacas Creek." Miller (1955a:163) heard the Barn Owl at 5000 feet in the oak belt on April 25 in the Sierra del Carmen.

****Otus asio suttoni*** Moore.--*O. a. suttoni* is found in the higher country of the Sierra del Carmen and western Coahuila. Miller (1955a:163) stated that Screech Owls were common in groves of oaks both at 7000 feet and 5000 feet in the Sierra del Carmen, and remarked that his series from the Sierra del Carmen matched well a series of *suttoni* from Chihuahua and Durango. Miller (1955a:163-164) also stated that Marsh took an adult at Jardin del Sur on August 28 and said that the specimen from the Sierra del Carmen referred to as *cineraceus* by Marsh and Stevenson (1938:286) agreed well with his series of *suttoni*. *O. a. suttoni* probably occurs no farther east than the Sierra del Carmen.

Otus asio mccallii* (Cassin).--*Specimens examined:* total 2: 32041 from 2 mi. W Jimenez, 850 ft., June 20, 1952; and sex ? 31645 from La Gacha, 1600 ft., December 9, 1953.

Three subspecies of the Screech Owl, *Otus asio*, occur in northeastern Mexico; two of these, *suttoni* and *mccallii*, occur in Coahuila, the latter in the eastern part. The third subspecies, *O. a. semplei*, occurs still farther east, for example in the state of Nuevo Leon on the Mesa del Chipinque 6 miles south of Monterrey (Sutton and Burleigh, 1939b:174).

Nos. 31645 and 32041 differ from *semplei* in that the tops of their heads do not appear to be solid blackish brown at a distance of four to five feet and in that the dark streakings of their backs and scapulars are not so heavy as in *semplei*. The mentioned specimens are brownish, not more black and white throughout as in *suttoni* nor are their toes heavily feathered (see Moore, 1941:154).

Findley observed a Screech Owl 2 mi. S and 3 mi. E San Juan de Sabinas on June 22, 1952. La Gacha would seem to represent the western extent of *mccallii* in Coahuila. *O. a. mccallii* and *suttoni* probably intergrade along the eastern slope of the Sierra del Carmen. Tordoff took No. 32041 near a tree that contained three young Screech Owls.

*****Otus flammeolus flammeolus*** (Kaup).--*Specimens examined:* total 2: 31600 from 20 mi. S Ocampo, 6000 ft., April 4, 1954; and 31581 from 13 mi. E San Antonio de las Alazanas, 9345 ft., April 9, 1954.

Miller (1955a:163) collected seven Flammulated Owls in the pines and oaks at 7000 feet in Carboneras Canyon and said that these owls were common there. Nos. 31600 and 31581 are suffused with cinnamoneous pigmentation, but represent the grayish phase, as described by Ridgway (1914:729). Van Hoose (1955:302) previously recorded Nos. 31600 and 31581 from Coahuila.

*****Bubo virginianus pallescens*** Stone.--*Specimens examined:* total 2: 32042 from 2 mi. S, 12 mi. E Nava, June 15, 1952; measurements: wing, 367 mm.; tail, 233 mm.; culmen, 29 mm.; and 31677 from 1.5 mi. NE Las Margaritas, 3100 ft., May 31, 1954; measurements: wing, 345 mm.; tail, 213 mm.; culmen, 26 mm.; testes, 8 mm. long.

Miller (1955a:164) took a male Great Horned Owl in the Sierra del Carmen on April 22. Ridgway (1914:742) listed *B. v. pallescens* from Sabinas. Sutton and Burleigh (1939a:30) obtained a female "near San Pedro" on January 29. Friedmann, Griscom, and Moore (1950:143) recorded *B. v. pallescens* from Coahuila on February 24, May 26, and June 10. The record of *B. v. mayensis* from Las Delicias (Amadon and Phillips, 1947:578) has been reidentified by Webster and Orr (1958:141) as *B. v. pallescens*. Dickerman saw a Great Horned Owl in the Sierra del Pino on May 12,1954.

****Glaucidium gnoma californicum*** Sclater.--*Specimen examined:* one, 31582, from 20 mi. S Ocampo, 6500 ft., April 5, 1954, weight, 55 gms.

Concerning forms of Pygmy Owls, Miller (1955a:164) remarked that the best distinguishing characters of *G. g. gnoma* and *californicum* are tail length and wing length. The characters of No. 31582 (wing, 94 mm.; tail, 69 mm.) are clearly those of *californicum* and not those of *gnoma*. Miller (*loc. cit.*) remarked that he "encountered at least five different individuals, chiefly in the pine-oak at 7000 feet" and one in oaks and pinons at 5000 feet in Boquillas Canyon of the Sierra del Carmen. The size of the testes (left, 9x5 mm.; right, 7x4 mm.) of No. 31582 indicates good probability of breeding by the Pygmy Owl in the State.

*****Micrathene whitneyi*** (Cooper).--Miller (1955a:164) heard the Elf Owl at close range in oaks at 5000 feet in Boquillas Canyon of the Sierra del Carmen on April 24.

*****Speotyto cunicularia hypugaea*** (Bonaparte).--*Specimens examined:* total 3: 32043 from 3 mi. S, 9 mi. E Cuatro Cienegas, 2250 ft., June 30, 1952; 32653 from 6 mi. W San Antonio de las Alazanas, July 5, 1955; and [** Male] 31602 from 14 mi. W San Antonio de las Alazanas, 6500 ft., January 7, 1954.

Although there are no records in Coahuila of specimens of Burrowing Owls north of 3 mi. S and 9 mi. E Cuatro Cienegas, this owl probably occurs in the northern section of the State. The records of Burrowing Owls from the southern part of Brewster County, Texas (Van Tyne and Sutton, 1937:38), suggest such occurrence. No. 31602 was shot in a prairie dog colony; No. 32043 was captured in a steel trap. Baker saw Burrowing Owls 7 mi. S and 4 mi. E Bella Union, 7200 feet, on June 25, 1952.

****Caprimulgus vociferus arizonae*** (Brewster).--*Specimens examined:* total 4: 31449 and 31450 from Sierra del Pino (=5 mi. S, 3 mi. W Acebuches), May 12 and 15, 1954, weights, 48 and 42 gms.; and 31028-31029 from 2 mi. N, 18 mi. W Santa Teresa, 7250 ft., April 3, 1952.

The Whip-poor-will occurs between 5000 feet and 9000 feet in Coahuila. Miller (1955a:164) reported *C. v. arizonae* in the Sierra del Carmen. Ridgway (1914:521) stated that *Antrostomus vociferus macromystax* (=*C. v. arizonae*) occurs in the southeastern sector of Coahuila at Sierra Guadalupe. Burleigh and Lowery (1942:190) reported the Whip-poor-will "near the summit of Diamante Pass"; because of its size (wing, 170 mm.; tail, 135 mm.) this Whip-poor-will from Diamante Pass seems to represent the subspecies *arizonae*. Dickerman saw four Whip-poor-wills 20 mi. S Ocampo, 6000 feet, on April 4, 1954. The sizes of the testes of Nos. 31449, 31028, and 31029 (13, 12, and 13 mm. long) and an egg taken from No. 31450 indicate breeding by this species in the State.

***Phalaenoptilus nuttallii nuttallii** (Audubon).--*Specimens examined:* total 6: 31032 from 37 mi. S, 21 mi. E Boquillas, 4100 ft., March 12, 1952; 31446 and 31447 from Sierra del Pino (=5 mi. S, 3 mi. W Acebuches), 6200 ft., May 13, 1954; 32048-32049 from 2 mi. S, 11 mi. E Nava, June 16, 1952; and 31033 from 4 mi. W Hacienda La Mariposa, 2300 ft., March 26, 1952.

Miller (1955a:164) found the Poor-will common along the rocky canyon walls up to 5000 feet in the Sierra del Carmen. Ridgway (1914:550) recorded the Poor-will at Sabinas on May 21 and at Saltillo on May 6. Van Tyne and Sutton (1937:39) stated that *P. n. nuttallii* was not common in the Big Bend Country of Texas; this probably is true for northwestern Coahuila as well. The specimens from the Sierra del Pino, collected by Dickerman in a pine-oak association at 6200 feet, were taken near the upper limit of their range. Findley saw Poor-wills 2 mi. W Jimenez, 850 feet, on June 19, 1952; 2 mi. S and 11 mi. E Nava, 810 feet, on June 15, 1952; and 2 mi. S and 3 mi. E San Juan de Sabinas on June 22, 1952. Dickerman noted Poor-wills 16 mi. E and 18 mi. N Ocampo on May 7, 1954, and 20 mi. S Ocampo, 6000 feet, on April 4, 1954.

The size of the testes (9x5 mm.) of No. 31446 and an enlarged oviduct and an ovum (4 mm. in diameter) of No. 31447 indicate breeding by the Poor-will in Coahuila.

****Chordeiles minor howelli** Oberholser.--*Specimen examined:* one, 31443, from 3 mi. N, 4 mi. E San Francisco (=25 mi. N Ocampo), 4850 ft., May 16, 1954.

Van Hoose (1955:302) wrote that nighthawks were heard and seen frequently 3 mi. N and 4 mi. E San Francisco. Blake (1953:227) said that the Common Nighthawk breeds in Sonora, Chihuahua, Tamaulipas, and Durango; the size of the testes (7x4.5 mm.) of No. 31443 indicates the possibility of breeding by this species in the State.

***Chordeiles acutipennis texensis** Lawrence.--*Specimens examined:* total 4: 32044 and 32045 from 2 mi. W Jimenez, 850 ft., June 20, 1952; 31442 from 5 mi. N, 13 mi. E Ocampo, May 6, 1954; and 32046 from 2 mi. N, 1 mi. W Ocampo, 4050 ft., July 6, 1952.

Specimens of the Lesser Nighthawk, subspecies *C. a. texensis*, have been recorded in the literature from representative localities throughout most of Coahuila. Burleigh and Lowery (1942:190-191) obtained, on April 18, at "a small pond at the edge of Saltillo," one male that was exceedingly fat; they (*loc. cit.*) suggested that their specimen was a migrant. Goldman (1951:377, 389) stated that *C. a. texensis* occupied the Lower and Upper Sonoran and Upper Austral life-zones of Coahuila. Dickerman saw Lesser Nighthawks at San Marcos (=20 mi. S Cuatro Cienegas) on May 4, 1954. Van Tyne and Sutton (1937:41) reported that the Lesser Nighthawk was common throughout the lower parts of the Big Bend in Texas. This is probably true for northwestern Coahuila as well.

The presence of an egg in the oviduct of No. 32045 and the dates (May 6, June 20, and July 6) on which Nos. 31442, and 32044-32046 were obtained indicate breeding by this species in Coahuila.

****Aeronautes saxatalis saxatalis** (Woodhouse).--*Specimen examined:* one, 31672, from Pico de Jimulco, 5600 ft., April 5, 1953, weight, 35 gms.

Burleigh and Lowery (1942:191) found the White-throated Swift to be common at "the summit of Diamante Pass and on the nearby ridges." Miller (1955a:164) saw the species from 4800 feet up to the crest of the Sierra del Carmen. Several White-throated Swifts were seen flying overhead at Pico de Jimulco on April 5.

No specimens of *A. s. sclateri* from Mexico are known. Miller (1955a:165) listed one specimen with dimensions (wing, 145 mm.) that approaches *sclateri*. The measurements of No. 31672 (wing, 143 mm.; tail, 58 mm.) also approach the dimensions of specimens of *sclateri* but are best referred to *A. s. saxatalis*.

****Calothorax lucifer** (Swainson).--Burleigh and Lowery (1942:191) obtained a male Lucifer Hummingbird at the Chorro del Agua on April 19. Van Tyne and Sutton (1937:43) reported a male from the Rio Grande (=3 mi. W Boquillas, Texas).

Archilochus colubris (Linnaeus).--Friedmann, Griscom, and Moore (1950:180) remarked that the Ruby-throated Hummingbird is a moderately common migrant, wintering from sea level to 9350 feet throughout Mexico, except in a few states. The only published record of a specimen of this hummer in the State is of a male taken on April 22 in a small arroyo twenty miles west of Saltillo (Burleigh and Lowery, 1942:191).

***Archilochus alexandri** (Bourcier and Mulsant).--*Specimens examined:* total 2: 31035 from the Rio Grande (=17 mi. S Dryden, Terrell County,

Texas, in Coahuila), 600 ft., March 18, 1952; and 32052 from 2 mi. S, 11 mi. E Nava, 810 ft., June 16, 1952.

Miller (1955a:165) stated the Black-chinned Hummingbird was common in the desert area at the base of the mountains of the Sierra del Carmen, and that Marsh, on July 25, obtained this hummingbird "near Piedra Blanca (Conejo)." Burleigh and Lowery (1942:191) obtained a female in an arroyo about twenty miles west of Saltillo on April 22.

*******Selasphorus platycercus platycercus*** (Swainson).--*Specimens examined:* total 2: 31583 from 20 mi. S Ocampo, 6500 ft., April 4, 1954; and 32673 from 13 mi. E San Antonio de las Alazanas, 9950 ft., July 6, 1955.

Miller (1955a:165) indicated that the Broad-tailed Hummingbird was fairly common in the Sierra del Carmen. According to Burleigh and Lowery (1942:191) this species is not uncommon in the pine woods bordering the summit of Diamante Pass; they saw the bird between 4900 and 10,000 feet.

****Eugenes fulgens fulgens*** Boucard.--Miller (1955a:165) obtained a male Rivoli's Hummingbird on Loomis Peak, 8800 feet, on April 11 in the Sierra del Carmen.

*****Lampornis clemenciae clemenciae*** (Lesson).--*Specimens examined:* total 2: 31036 from 26 mi. W Santa Teresa, 7050 ft., April 5, 1952; and 32668 from 13 mi. E San Antonio de las Alazanas, 9950 ft., July 6, 1955.

The Blue-throated Hummingbird occurs between 5000 and 9950 feet in Coahuila. Miller (1955a:165) remarked that it was found in canyon bottoms, "whether at 7500 feet among the rocky slopes, oaks, and white pines ... or at 5000 feet in the madrone, maples, elms, and basswoods...." No. 32668, was obtained in a Douglas fir-pine-aspen association at 9950 feet.

[***Amazilia yucatanensis chalconota*** Oberholser.--The Buff-bellied Hummingbird seems to be uncommon in Coahuila. Friedmann, Griscom, and Moore (1950:172) and the AOU Check-list Committee (1957:306) stated that this hummingbird occurs in Coahuila.]

Megaceryle alcyon alcyon (Linnaeus).--Miller (1955a:165) saw a Belted Kingfisher at Carboneras Canyon in the Sierra del Carmen. Miller, Friedmann, Griscom, and Moore (1957:15) stated that one specimen of *M. a. alcyon* was obtained in Coahuila on November 14.

*****Chloroceryle americana hachisukai*** (Laubmann).--*Specimens examined:* total 2: 32053 from 12 mi. N, 12 mi. W Jimenez, June 19, 1952; and 31038 from 8 mi. N, 4 mi. W Muzquiz, 1800 ft., March 31, 1952.

The Green Kingfisher has been recorded in Coahuila as far south as 8 mi. N and 4 mi. W Muzquiz. The forehead of No. 31038, when compared with typical representatives of *C. a. hachisukai,* is not extensively streaked with white, nor are all the coverts conspicuously spotted with white, yet it clearly has more extensive white markings than typical representatives of *C. a. septentrionalis*. This specimen from 8 mi. N and 4 mi. W Muzquiz probably is intermediate between *hachisukai* and *septentrionalis*. Miller (1955a:165) stated that Marsh took a specimen at Tanque de los Melones on La Bavia Ranch east of Fresno Mesa in the Sierra del Carmen that is a typical *C. a. hachisukai.*

Colaptes cafer collaris Vigors.--Ridgway (1914:34) listed this subspecies of the Red-shafted Flicker from Saltillo and Agua Nueva.

****Colaptes cafer nanus*** Griscom.--*Specimens examined:* total 3: 31463 from Sierra del Pino (=5 mi. W, 3 mi. S Acebuches), May 13, 1954, weight, 108 gms.; and 31651 and 31652 from the north foot of Sierra Guadalupe (=10 mi. S, 5 mi. W General Cepeda), 6500 ft., April 21, 1953.

In suitable habitats in Coahuila the Red-shafted Flicker is common. Miller (1955a:165-166) stated that *C. c. nanus* was common at Corte Madera Canyon, Boquillas Canyon, and Carboneras Canyon in the Sierra del Carmen and recorded a specimen also from Sierra de Jardin on August 7. Burleigh and Lowery (1942:192) recorded *C. c. nanus* from Diamante Pass and Saltillo. Sutton and Burleigh (1939a:31) stated that the Red-shafted Flickers, obtained 5 mi. E La Rosa and at Diamante Pass, may be intermediate between *C. c. collaris* (then called *C. c. chihuahuae*) and *C. c. nanus*. Nesting of the Red-shafted Flicker in Coahuila was reported by Miller (1955a:165) and Burleigh and Lowery (1942:192). Hardy saw Red-shafted Flickers 13 mi. E San Antonio de las Alazanas on July 6, 1955.

****Centurus aurifrons aurifrons*** (Wagler).--*Specimens examined:* total 6: 32054 from 12 mi. N, 12 mi. W Jimenez, 850 ft., June 19, 1952; 32055-32057 from 2 mi. W Jimenez, 850 ft., June 20, 1952; 31039 from 8 mi. N, 4 mi. W Muzquiz, 1800 ft., April 1, 1952; and 33150 (skeleton only) from Parras, July 4, 1955.

The Golden-fronted Woodpecker occurs throughout Coahuila, but uncommonly in the western part. Wetmore (1948:185-186) examined a series of Golden-fronted Woodpeckers that showed a definite cline in dorsal coloration from north-central Texas to Jalisco, Michoacan, and Hidalgo in south-central Mexico. He (*loc. cit.*) stated that "the extremes are easily separable, but in southern and southwestern Texas and in northeastern Mexico the two kinds [*C. a. aurifrons* and *C. a. incanescens*] ... merge so gradually that over a broad area the whole population is intermediate, making decisions as to any sharply drawn dividing line difficult and in part arbitrary." *C. a. incanescens,* according to Wetmore, occurs in western and central Texas south to northeastern Chihuahua and northern Coahuila whereas *C. a. aurifrons* occurs in north-central Coahuila (Monclova) and southern Texas south to Jalisco, Michoacan, Hidalgo, and central Tamaulipas.

The specimens that I have examined from Coahuila are variably intermediate between the subspecies *aurifrons* and *incanescens*. The dark and white cross-bars on the back of No. 31039 are nearly equal (dark bands wider in *aurifrons*; white bands wider in *incanescens*); the rump and upper tail coverts are more or less mixed with black as in *aurifrons*. The dark cross-bars on the backs of Nos. 32054-32057 are slightly larger than the white cross-bars; the rump and upper tail coverts of these specimens are somewhat mixed with black.

Centurus aurifrons incanescens* Todd.--This subspecies of the Golden-fronted Woodpecker is listed by Miller, Friedmann, Griscom, and Moore (1957:34) from "... northern Coahuila (upper Rio Grande valley)." Burleigh and Lowery (1942:192) stated that the Golden-fronted Woodpecker "apparently avoids the mountain slopes, but was found to be not uncommon on the arid plateau about Saltillo." Findley saw Golden-fronted Woodpeckers 2 mi. W Jimenez, 850 feet, on June 19, 1952; and 2 mi. S and 3 mi. E San Juan de Sabinas on June 22, 1952. Cory (1919:424) listed the Golden-fronted Woodpecker from Sabinas. Nos. 32055-32057 are juveniles, and thus document breeding by this woodpecker in Coahuila.

Melanerpes formicivorus formicivorus* (Swainson).--*Specimens examined:* total 5: 31040 (skeleton only) from Fortin (=33 mi. N, 8 mi. W San Geronimo), 3300 ft., March 28, 1952; 29423 and 29424 from Club Sierra del Carmen (=2 mi. N, 6 mi. W Piedra Blanca), 4950 ft., April 8, 1950; 31041 (skeleton only) from 26 mi. W Santa Teresa, 7050 ft., April 5, 1952; and 31668 from Sierra Guadalupe, Canon d. Meco (=10 mi. S General Cepeda), 6500 ft., April 23, 1953.

The Acorn Woodpecker in Coahuila is common in the oak and pine-oak belts, from 4950 to 8000 feet. Miller (1955a:166) stated that the Acorn Woodpecker in the Sierra del Carmen was an abundant and conspicuous bird throughout the oak and pine-oak belts, from 5000 to 8000 feet. Dickerman saw two Acorn Woodpeckers in the Sierra de la Madera on December 13, 1953, and four 20 mi. S Ocampo, 6000 feet, on April 4, 1954. Breeding of *M. f. formicivorus* was reported by Miller (*loc. cit.*) who took females nearly ready to begin laying; one of our specimens (No. 29423) had enlarged testes (11 mm.).

Sphyrapicus varius varius (Linnaeus).--*Specimen examined:* one, 31649, from the north foot of Sierra Guadalupe (=10 mi. S, 5 mi. W General Cepeda), 6400 ft., April 19, 1953.

Miller (1955a:166) reported this Yellow-bellied Sapsucker as a winter visitant or migrant in the Sierra del Carmen. Burleigh and Lowery (1942:192) took a female *S. v. varius* at the summit of Diamante Pass on April 14. Ridgway (1914:275) listed *S. v. varius* from Sierra de Guadalupe.

Sphyrapicus varius nuchalis Baird.--Miller (1955a:166) reported this Yellow-bellied Sapsucker as "indeed common" in the Sierra del Carmen, and indicated that both *S. v. nuchalis* and *S. v. varius* were "found only at the upper levels in the pine-oak formation and usually in relatively dense clumps of trees in the canyon bottoms." Ridgway (1914:280) listed *C. v. nuchalis* from Rio Sabinas.

****Dendrocopos villosus icastus** (Oberholser).--Miller, Friedmann, Griscom, and Moore (1957:39) remarked that this subspecies of the Hairy Woodpecker occurs, in northwestern Mexico, from eastern Sonora, Jalisco, Chihuahua, Durango, Zacatecas to southern Coahuila. Ridgway (1914:222) stated that *D. v. icastus* occurs in Coahuila at Carneros and Sierra de Guadalupe, the former being the easternmost known locality for the subspecies.

****Dendrocopos villosus intermedius** (Nelson).--*Specimens examined:* total 2: 32701 from 13 mi. E San Antonio de las Alazanas, 9950 ft., July 6, 1955; and 31604 from 2 mi. E Mesa de Tablas, 8500 ft., January 15, 1954.

In northeastern Mexico this subspecies, according to Miller, Friedmann, Griscom, and Moore (1957:39), occurs in Guanajuato, San Luis Potosi, Hidalgo, Nuevo Leon, and Tamaulipas. Nos. 31604 and 32701 represent the first records of *D. v. intermedius* from Coahuila.

***Dendrocopos scalaris cactophilus** (Oberholser).--*Specimens examined:* total 2: 31042 from 7 mi. S, 2 mi. E Boquillas, 800 ft., March 1, 1952; and 31043 from 10 mi. S, 5 mi. E Boquillas, 1500 ft., March 3, 1952.

These specimens of the Ladder-backed Woodpecker show signs of intergradation with *D. s. symplectus*. Both specimens are pale enough above for *D. s. symplectus*, but the wing and the tail of each (wing, 102, 103 mm.; tail, 60, 65 mm.) are too long for *symplectus*. This suggestion of intergradation is not unexpected because *symplectus* occurs in northeastern Coahuila and *cactophilus* in the Chisos Mountains of Texas. Miller (1955a:166) also obtained one female *D. s. cactophilus* from the Sierra del Carmen that suggested intergradation with *symplectus*.

Miller (*loc. cit.*) wrote that *D. c. cactophilus* was found chiefly in the oaks and was common in the lower oak belt at 5000 feet. The upper limit of the range of the Ladder-backed Woodpecker, according to Miller, is 6800 feet. He reported the species as breeding in the Sierra del Carmen.

Miller (1955b:317) took a hybrid woodpecker representing a cross between *Dendrocopos villosus* and *Dendrocopos scalaris* in the Sierra del Carmen, where, although Ladder-backed Woodpeckers were common, he found no Hairy Woodpeckers.

*****Dendrocopos scalaris symplectus*** (Oberholser).--*Specimens examined:* total 2: 32058 from 2 mi. W Jimenez, June 20, 1952; and 31667 from Sierra Guadalupe, Domingo Canyon (=10 mi. S, 14 mi. W General Cepeda), 6700 ft., April 18, 1953.

This Ladder-backed Woodpecker, according to the AOU Check-list Committee (1957:327) and Oberholser (1912:156), occurs in Texas (east of Pecos), northeastern Coahuila, Nuevo Leon, and Tamaulipas. The area of intergradation of *D. s. symplectus* and *giraudi* is in southeastern Coahuila. The dark smoky underparts and the equal size of the white and black bars of the upper parts of No. 31667 suggest intergradation with *D. s. giraudi*. Yet, the size of the wing indicates that this specimen is closer to *D. s. symplectus*. No. 32058 has characters of typical representatives of *D. s. symplectus*. Burleigh and Lowery (1942:193) reported *D. s. symplectus* "in the desert country west of Saltillo." Ridgway (1914:257) listed the Ladder-backed Woodpecker from Sabinas. Cory (1919:494) listed *D. s. symplectus* from Pabinas (=Sabinas?).

*****Dendrocopos scalaris giraudi*** (Stone).--*Specimen examined:* one, 34623, from 12 mi. N, 10 mi. E Parras, 3850 ft., July 12, 1953, weight, 35 gms.

Ridgway (1914:259) reported that *D. s. giraudi* (then called *Dryobates scalaris bairdi*) has been recorded from La Ventura, Carneros, and Jaral. Oberholser (1912:159) indicated that the subspecies *giraudi* occurs north to Jaral in southern Coahuila, east to La Ventura in southeastern Coahuila, and south through central Mexico. The present specimen is darker ventrally than specimens of either *D. s. cactophilus* or *D. s. symplectus,* and on its upper surface the black bars are wider than the white. The specimen is in worn plumage, but nevertheless suggests intergradation with *D. s. symplectus*.

Platypsaris aglaiae albiventris (Lawrence).--The Rose-throated Becard, if it occurs in Coahuila at all, is rare in the State. Without giving any specific locality, Sclater (1857:74) described *Pachyrhamphus aglaiae* from Coahuila. Hellmayr (1929:202) considered Sclater's record as representing *Platypsaris aglaiae albiventris* (Lawrence).

*****Tyrannus vociferans vociferans*** (Swainson).--*Specimens examined:* total 2: 32064 from 7 mi. S, 4 mi. E Bella Union, 7200 ft., June 24, 1952; and 31650 from the north foot of Sierra Guadalupe (=10 mi. S, 5 mi. W General Cepeda), 6400 ft., April 15, 1953.

The small number of records of Cassin's Kingbird in Coahuila is surprising, for I would expect the species in most areas of the State between 6000 and 7000 feet. All Coahuilan records are from the southeastern part of the State. Burleigh and Lowery (1942:193) found it "on infrequent occasions in the arid country near Saltillo." Miller, Friedmann, Griscom, and Moore (1957:71) reported that *T. v. vociferans* nested in Coahuila on July 7. The size of the testes (15x7 mm.) of our specimen from near Bella Union also indicates breeding.

*****Muscivora forficata*** (Gmelin).--*Specimen examined:* one, 32063, from 12 mi. N, 12 mi. W Jimenez, 850 ft., June 19, 1952.

The status of the Scissor-tailed Flycatcher in Coahuila is uncertain. Although the condition of the gonads of No. 32063 was not recorded by the collector, the late date (June 19) on which is was obtained suggests that this female was a resident in northeastern Coahuila. Findley saw a Scissor-tailed Flycatcher 2 mi. S and 11 mi. E Nava, 810 feet, on June 15, 1952. Dickerman saw one 4 mi. N San Isidro on May 10, 1954. Miller, Friedmann, Griscom, and Moore (1957:69) recorded this flycatcher from Sabinas on April 12. No other records of *Muscivora forficata* in Coahuila have come to my attention, but the species probably occurs in the State in migration.

****Myiarchus crinitus crinitus*** (Linnaeus).--*Specimens examined:* total 3: 32065 from 12 mi. N, 12 mi. W Jimenez, 850 ft., June 19, 1952; and 32066 and 32067 from 2 mi. W Jimenez, 850 ft., June 20, 1952.

The Great Crested Flycatcher seems to be rare in Coahuila. Nos. 32065-32067 are the first specimens that I know of from Coahuila. Bangs (1898:179-180) said that the subspecies *crinitus* has an "enormous swollen bill" (exposed culmen of male, 20 mm.; breadth of bill at middle of nostril, 10.3 mm.) whereas *M. c. boreus* has a "small slender bill" (exposed culmen of male, 18.6 mm.; breadth of bill at middle of nostril, 8.6 mm.). No. 32066 has a large bill (exposed culmen, 21 mm.; breadth of bill at middle of nostril, 10 mm.). Bangs (*loc. cit.*) did not give any measurements for females of *M. c. crinitus* or *boreus*. Nevertheless, I suspect that Nos. 32065 and 32067 represent *crinitus* (No. 32065: exposed culmen, 19 mm.; breadth of bill at middle of nostril, 9 mm.; No.

32067: exposed culmen, 20 mm.; breadth of bill at middle of nostril, 9.5 mm.). The size of the testes (9 mm. long) of No. 32066 and the dates (June 19 and 20) on which the specimens were collected indicate breeding by this species in the State.

****Myiarchus tyrannulus cooperi*** Baird.--*Specimens examined:* total 2: 32068 and 32069 from 2 mi. S, 3 mi. E San Juan de Sabinas, 1160 ft., June 22, 1952, measurements: wing, 102, 97 mm.; tail, 93, 90 mm.

Wied's Crested Flycatcher is not common in Coahuila. Eastern Coahuila represents the eastern limit of the range of *M. t. cooperi.* Measurements of our specimens agree well with the average measurements of typical *M. t. cooperi.* According to Ridgway (1907:621), *M. t. cooperi* (then called *Myiarchus mexicanus mexicanus*) has been recorded from Sabinas. The size of the testes (14x7 mm.) of No. 32068 and the date (June 22) on which the specimens were collected indicate breeding by this species in the State.

****Myiarchus cinerascens cinerascens*** (Lawrence).--*Specimens examined:* total 7: 31045 from 15 mi. SE Boquillas, 1500 ft., March 16, 1952; 32070 from 12 mi. N, 12 mi. W Jimenez, 850 ft., June 19, 1952; 32071 from 2 mi. W Jimenez, 850 ft., June 20, 1952; 35413 from 6 mi. N, 2 mi. W Castillon, 3750 ft., June 30, 1953; 32072 from 9 mi. S, 11 mi. E Sabinas, June 14, 1952; 31584 from 17 mi. S Ocampo, 5300 ft., April 7, 1954; and 31673 from the west foot of Pico de Jimulco, 5000 ft., April 4, 1953.

The Ash-throated Flycatcher is common in Coahuila. All specimens examined by me from there are typical of *M. c. cinerascens.* Ridgway (1907:626) listed *M. c. cinerascens* from Monclovia (=Monclova?) and Sabinas. Miller (1955a:166) found the species "only in the oak belt at 5000 feet, where it was common." Burleigh and Lowery (1942:193) recorded *M. c. cinerascens* from "near Saltillo." Dickerman obtained No. 31584 in a mesquite-grassland-shrubby area. Miller, Friedmann, Griscom, and Moore (1957:78) stated that *M. c. cinerascens* breeds in Coahuila.

Sayornis phoebe (Latham).--Miller, Friedmann, Griscom, and Moore (1957:66) recorded the Eastern Phoebe from Coahuila on November 4-14.

****Sayornis nigricans semiatra*** (Vigors).--*Specimens examined:* total 2: 31046 from 1 mi. N Boquillas, 700 ft., March 7, 1952; and 31047 from Fortin (=33 mi. N, 8 mi. W San Geronimo), 3300 ft., March 27, 1952.

Of the Black Phoebe, the two subspecies *semiatra* and *nigricans* intergrade in Coahuila. Typical representatives of *S. n. semiatra* are present in northern Coahuila. The under tail coverts of Nos. 31046-31047 are immaculate and white. Miller (1955:167) noted the two specimens collected from the Sierra del Carmen to have narrow dark shaft streaks on the under tail coverts. He (*loc. cit.*) remarked also that "the marking of the under tail coverts may indicate a beginning of a gradient in increased darkening of these feathers toward *S. n. nigricans* in southern Coahuila."

Sayornis nigricans nigricans* (Swainson).--Burleigh and Lowery (1942:193) collected several specimens of *S. n. nigricans* "on the outskirts of Saltillo" and saw a pair at the Chorro del Agua on April 19.

****Sayornis saya saya*** (Bonaparte).--*Specimens examined:* total 4: 31049 from the Rio Grande (=17 mi. S Dryden, Terrell Co., Texas, in Coahuila), 600 ft., March 18, 1952; sex ? 31048 from 10 mi. S, 5 mi. E Boquillas, 1500 ft., March 4, 1952; 31050 from Fortin (=33 mi. N, 8 mi. W San Geronimo), 3300 ft., March 29, 1952; and 32059 from 7 mi. S, 4 mi. E Bella Union, 7200 ft., June 25, 1952.

Say's Phoebe occurs commonly in Coahuila. Miller (1955a:167) obtained a female with an active brood patch in Boquillas Canyon of the Sierra del Carmen, and remarked that Marsh took a juvenile on September 2 at El Rincon. Sutton and Burleigh (1939a:33) saw this phoebe several times in southern Coahuila and obtained a male "near San Pedro" on January 29. Burleigh and Lowery (1942:193) recorded this species as "a common breeding bird both on Diamante Pass and on the arid plateau around Saltillo." No. 32059 was a juvenile.

Empidonax traillii brewsteri Oberholser.--Amadon and Phillips (1947:578) look a Traill Flycatcher of the subspecies *brewsteri* at Las Delicias on August 11.

Empidonax minimus (Baird and Baird).--*Specimen examined:* one, 31470, from Sierra del Pino (=5 mi. S, 3 mi. W Acebuches), 6200 ft., May 13, 1954.

Amadon and Phillips (1947:578) obtained two Least Flycatchers at Las Delicias on August 12. Dickerman took No. 31470 in pine-oak vegetation.

Empidonax hammondii (Xantus).--*Specimen examined:* one, sex ? 31657, from the north slope of Sierra Guadalupe (=11 mi. S, 7 mi. W General Cepeda), 7800 ft., April 20, 1953.

No. 31657 is similar to *E. wrightii* (Wright's Flycatcher); however, the outmost (tenth) primary is equal to or slightly larger than the fifth primary. Yet, the underparts of No. 31657 are darker and more uniform in coloration than those of typical representatives of *E. wrightii*. Miller, Friedmann, Griscom, and Moore (1957:88) stated that Hammond's Flycatcher is "transient" in Coahuila. Burleigh and Lowery (1942:193-194) reported that *E. hammondii* was the most prevalent of the small flycatchers in southeastern Coahuila. They (*loc. cit.*) obtained specimens of this flycatcher from the Chorro del Agua and Diamante Pass. Miller

(1955a:167) characterized *E. hammondii* as a common migrant, "chiefly in stands of low oaks in the pine-oak belt but also occasionally in the desert scrub" of the Sierra del Carmen.

Empidonax wrightii Baird.--Amadon and Phillips (1947:578) reported one Wright's Flycatcher from Las Delicias. Miller (1955a:167) stated that this species was a common migrant and occurred chiefly in the lower oak belt and in the desert scrub. Sutton and Burleigh (1939a:34) obtained specimens of *E. wrightii* from San Pedro on January 29.

Empidonax griseus Brewster.--Sutton and Burleigh (1939a:34) noted that the Gray Flycatcher was common "in the San Pedro region" and collected two at San Pedro on January 29. Burleigh and Lowery (1942:194) collected specimens "near the Chorro del Agua, at Saltillo, and ... in the open desert some twenty miles west of Saltillo."

****Empidonax affinis trepidus** Nelson.--*Specimen examined:* one, 32750, from 13 mi. E San Antonio de las Alazanas, 9950 ft., July 6, 1955.

Ridgway (1907:576) previously reported the little known and poorly defined Pine Flycatcher from Sierra Guadalupe. Because of its small size (wing, 75 mm.; tail, 65 mm.), No. 32750 is referable to *E. a. trepidus*. No indication of breeding of the subspecies *trepidus* exists for Coahuila. Nevertheless, the date (July 6) on which No. 32750 was obtained suggests that this flycatcher may breed in southeastern Coahuila.

***Empidonax difficilis hellmayri** Brodkorb.--*Specimen examined:* one, 31469, from Sierra del Pino (=5 mi. S, 3 mi. W Acebuches), May 13, 1954, measurements: wing, 71 mm.; tail, 65 mm.; culmen, 11.5 mm.; tarsus, 17.0 mm.

Miller (1955a:167) reported that the Western Flycatcher breeds in the Sierra del Carmen, from 6800 to 7500 feet. According to Miller, Friedmann, Griscom, and Moore (1957:91), this subspecies has been recorded from Sierra Guadalupe. No. 31469 closely resembles, especially in measurements, the specimens of the Western Flycatcher from the Chisos Mountains of Texas as reported by Brodkorb (1935:2).

Empidonax difficilis subsp.--Burleigh and Lowery (1942:194) obtained a specimen of *E. difficilis* "near the summit of Diamante Pass" that they tentatively identified as *E. d. salvini*. However, they considered the specimen as possibly *E. d. immemoratus* or *E. d. occidentalis*.

Localities listed by Miller, Friedmann, Griscom, and Moore (1957:92) of *E. d. salvini* and *E. d. immemoratus* are south of twenty-three degrees north latitude whereas the range of *E. d. occidentalis* includes parts of Nuevo Leon north of twenty-three degrees north latitude (*occidentalis* intergrades with *hellmayri* at Cerro Potosi, Nuevo Leon, [Miller,

Friedmann, Griscom, and Moore, *loc. cit.*]). Thus, the specimen from Diamante Pass probably is either *E. d. hellmayri* or *E. d. occidentalis*.

*****Empidonax fulvifrons pygmaeus*** Coues.--Miller, Friedmann, Griscom, and Moore (1957:95) listed this subspecies of the Buff-breasted Flycatcher from the Sierra Guadalupe on April 21. This record represents the eastern limit of the range of *E. f. pygmaeus* in northeastern Mexico.

*****Contopus pertinax pertinax*** Cabanis and Heine.--Burleigh and Lowery (1942:194) obtained one female Coues' Flycatcher "in a small gully just below the summit of Diamante Pass."

Contopus virens (Linnaeus).--I judge from the paucity of records in the literature that the Eastern Wood Pewee is uncommon in Coahuila. Burleigh and Lowery (1942:194) obtained two males at the Chorro del Agua, and remarked also that "it is possible that as far west as Saltillo, this species is a rather uncommon transient." Ridgway (1907:519) listed *Contopus virens* from Sabinas.

****Contopus sordidulus veliei*** Coues.--*Specimens examined:* total 2: 31467 from Sierra del Pino (=5 mi. S, 3 mi. W Acebuches), May 14, 1954; and 31653 from the north foot of Sierra Guadalupe (=10 mi. S, 5 mi. W General Cepeda), 6400 ft., April 19, 1953.

Specimen No. 31467 of the Western Wood Pewee was obtained in pine and oak vegetation by Dickerman. Ridgway (1907:523) reported *Contopus sordidulus veliei,* under the name *Myiochanes richardsonii richardsonii,* from Sierra Encarnacion. Amadon and Phillips (1947:578) obtained a specimen of *C. sordidulus* from Las Delicias. Miller, Friedmann, Griscom, and Moore (1957:83) remarked that the Western Wood Pewee breeds in the State.

Nuttallornis borealis (Swainson).--Several records of the Olive-sided Flycatcher from Coahuila are present in the literature. Miller (1955a:167) reported it as a migrant in the desert at the base of the Sierra del Carmen on April 24. Miller, Friedmann, Griscom, and Moore (1957:82) reported this species from Ocampo on June 16. Amadon and Phillips (1947:578) obtained one at Las Delicias on August 16. Burleigh and Lowery (1942:194) collected two males "near the summit of Diamante Pass" on April 14.

****Pyrocephalus rubinus mexicanus*** Sclater.--*Specimens examined:* total 3: 32060 and 32061 from 12 mi. N, 12 mi. W Jimenez, 850 ft., June 19, 1952; and sex ? 32062 from 2 mi. W Jimenez, 850 ft., June 20, 1952.

The Vermilion Flycatcher is common in Coahuila. Except in the northwestern part of the State, the subspecies *mexicanus* is present throughout Coahuila. The size of No. 32060 (wing, 80 mm.; tail, 62 mm.)

suggests that the specimen is an intergrade between *P. r. flammeus* and *mexicanus*. *P. r. flammeus* and *mexicanus* seem to intergrade in northern Coahuila.

Burleigh and Lowery (1942:195) found *mexicanus* to be "quite plentiful on the plains surrounding Saltillo." Sutton and Burleigh (1939a:33) noted the Vermilion Flycatcher "near San Pedro." Cory and Hellmayr (1927:92) listed *P. r. mexicanus* from Sabinas. The size of the testes (6x4 mm. long) of No. 32060 and the dates (June 19 and 20) on which our specimens were collected indicate breeding.

Pyrocephalus rubinus flammeus van Rossem.--This subspecies of Vermilion Flycatcher occupies the northwestern section of Coahuila. Marsh and Stevenson (1938:287) took a specimen of *P. r. flammeus* "near Santo Domingo, east of the Del Carmens...." Miller (1955a:167) re-examined this specimen and stated that "its affinity with the northwest race of the species seems correct in terms of the characters outlined by van Rossem (1934:353)."

Eremophila alpestris enthymia (Oberholser).--Miller, Friedmann, Griscom, and Moore (1957:105) recorded this subspecies of Horned Lark wintering 4 mi. S Hipolito (November 2 to February 24). They reported also that *E. a. enthymia* breeds in Coahuila (4 mi. S Hipolito ?).

****Eremophila alpestris diaphora*** (Oberholser).--*Specimens examined:* total 9: 32073-32078 and 32079 from 7 mi. S, 4 mi. E Bella Union, 7200 ft., June 25, 1952; and 31605-31606 from 14 mi. W San Antonio de las Alazanas, 6500 ft., January 9, 1954.

Burleigh and Lowery (1942:195) wrote that this subspecies of Horned Lark was uncommon "about Saltillo," but "fairly common" in Diamante Valley, at about 7000 feet. Oberholser (1902:863) recorded the subspecies *aphrasta* from La Ventura. Ridgway (1907:326) listed *Otocoris alpestris aphrasta* Oberholser from Saltillo and La Ventura. However, Miller, Friedmann, Griscom, and Moore (1957:105) indicated that *E. a. diaphora* rather than *aphrasta* was recorded from Saltillo and La Ventura. Sutton and Burleigh (1939a:35) remarked that "specimens collected near Ramos Arizpe [in southeastern Coahuila]" proved to be *E. a. aphrasta*. However, this record might be questioned, as Miller, Friedmann, Griscom, and Moore (*op. cit.*) indicated, because *diaphora*, rather than *aphrasta*, seems to occupy the southeastern sector of Coahuila. Possibly typical representatives of *aphrasta* and/or intergrades between *aphrasta* and *diaphora* are present in western Coahuila; however, no records of *E. a. aphrasta* exist from western Coahuila.

The sizes of Nos. 32073-32077 and 31605-31606 (wing, 98.0-101.5 mm., averaging 99.7 mm.), the bright yellow throat, and the vinaceous color of the hindneck characterize clearly the subspecies *diaphora*. The sizes of the testes (9x6 mm.; 8x5 mm.; 10x6 mm.; 8x4 mm.; 11x6 mm.) of Nos. 32073-32077, the size of the largest ovum (6.5 mm.) of No. 32079, and the juvenile (32078) are evidence of breeding of *E. a. diaphora* in Coahuila.

Tachycineta thalassina thalassina* (Swainson).--*Specimens examined:* total 3: 31471, 31473, and 31472 from Sierra del Pino (=5 mi. S, 3 mi. W Acebuches), May 15, 1954, measurements: wings, 125, 118, 108 mm.; tails, 56, 54, 46 mm.

The two subspecies of the Violet-green Swallow, *thalassina* and *lepida*, intergrade in Coahuila. Specimens from different localities in the State represent various stages of intergradation between the two subspecies; generally those from northern Coahuila seem to be closer to *T. t. lepida*, and those from southern Coahuila are closer to *T. t. thalassina*. Nos. 31471-31473 are intergrades between *T. t. thalassina* and *T. t. lepida*; in size the three resemble *T. t. thalassina*, but in green, rather than purple, backs and scapulars resemble *T. t. lepida*. The rumps of Nos. 31471-31473 show some purple with the green, but are nearer *thalassina* in this character.

Gonadal sizes (testes 10x7, 10x8 mm., one ovum 6 mm.) indicate that the Violet-green Swallow breeds in the Sierra del Pino.

****Tachycineta thalassina lepida*** Mearns.--Miller (1955a:167) reported that Violet-green Swallows taken in Boquillas Canyon of the Sierra del Carmen are intermediate between *T. t. lepida* and *thalassina*; however, he referred his sample to *lepida* on the basis of short wing.

Iridoprocne bicolor (Vieillot).--Miller, Friedmann, Griscom, and Moore (1957:114) reported the Tree Swallow from Hipolito on February 22.

Stelgidopteryx ruficollis psammochrous Griscom.--Miller, Friedmann, Griscom, and Moore (1957:111) reported this subspecies of the Rough-winged Swallow from Saltillo.

****Hirundo rustica erythrogaster*** Boddaert.--Burleigh and Lowery (1942:195) stated that the Barn Swallow was the most abundant swallow "about Saltillo." Because these swallows occupied houses about Saltillo and neighboring villages, Burleigh and Lowery (*loc. cit.*) concluded that the species nests commonly in the Saltillo area. Findley saw Barn Swallows 2 mi. S and 11 mi. E Nava, 810 feet, on June 15, 1952. Dickerman observed them at San Marcos (=20 mi. S Cuatro Cienegas) on May 4, 1954.

****Petrochelidon pyrrhonota minima*** van Rossem and Hachisuka.--*Specimens examined:* total 2: 31585-31586 from 14 mi. W San Antonio de las Alazanas, April 9, 1954.

Burleigh and Lowery (1942:196) noted the Cliff Swallow "in small numbers about Saltillo and the nearby villages" where the species "showed evidence of beginning to nest in many of the houses...." They (*loc. cit.*) found Cliff Swallows also at the Chorro del Agua and Diamante Valley. According to Dickerman, Nos. 31585-31586 were from a colony, members of which were collecting mud at a stock tank.

****Petrochelidon fulva pallida*** Nelson.--The Cave Swallow seems to be uncommon in eastern Coahuila. Selander and Baker (1957:345) list Saltillo, Sabinas, and Monclova as the three known localities for this swallow in the State.

****Progne subis subis*** (Linnaeus).--Miller, Friedmann, Griscom, and Moore (1957:107) reported, presumably on the basis of a published record not found by me, that the Purple Martin breeds in Coahuila.

*****Progne chalybea chalybea*** (Gmelin).--The only record of the Gray-breasted Martin in Coahuila was given by Ridgway (1904:42) when he listed *P. c. chalybea* from Sabinas.

*****Cyanocitta stelleri macrolopha*** Baird.--*Specimens examined:* total 2: 32788 and 32787 from 13 mi. E San Antonio de las Alazanas, 9950 ft., July 6, 1955.

From the paucity of records in the literature, I judge that Steller's Jay is uncommon in Coahuila. Nos. 32787-32788 seemingly represent the first records of this species in the State, and are referred to the subspecies *macrolopha* on the basis of relatively long (150, 151 mm.) wing, near the maximum for *stelleri* from Mexico. The date (July 6) of collection suggests that these birds were resident in southeastern Coahuila.

*****Aphelocoma coerulescens cyanotis*** Ridgway.--The Scrub Jay is common in southeastern Coahuila. Burleigh and Lowery (1942:196) noted this species in small numbers at the summit of Diamante Pass, daily on the lower slopes of the mountains, and a few birds "on the outskirts of Saltillo, where they were probably nesting...." Sutton and Burleigh (1939a:35) also noted several flocks at Diamante Pass. Miller, Friedmann, Griscom, and Moore (1957:123) reported *A. c. cyanotis* from El Diamante. Ridgway (1904:335) listed this subspecies of the Scrub Jay (then called *Aphelocoma cyanotis*) from Carneros, Sierra Encarnacion, and Sierra Guadalupe.

****Aphelocoma ultramarina couchii*** (Baird).--*Specimens examined:* total 14: 31051 from Sierra de la Encantada (=38 mi. S, 23 mi. E Boquillas), 4400 ft., March 15, 1952; 29425-29426 and 29427-29428 from Club Sierra del Carmen (=2 mi. N, 6 mi. W Piedra Blanca), 4950 ft., April 8, 1950; sex ? 31052 (skeleton only) from 4 mi. W Hacienda La Mariposa, 2300 ft., March 25, 1952; 31635-31636 from Canyon de Parazos in the Sierra de la Parazos Azula (=9 mi. E Hermanas), 2100 ft., December 7, 1953; 32082 from 4 mi. N, 21 mi. W Cuatro Cienegas, 6200 ft., July 3, 1952; 31053 (skeleton only) from 26 mi. W Santa Teresa, 7050 ft., April 5, 1952; 32081 from 7 mi. S, 4 mi. E Bella Union, 7200 ft., June 25, 1952; 33173 (skeleton only) from 13 mi. E San Antonio de las Alazanas, 9950 ft., July 6, 1955; 31607 from 13 mi. E, 3 mi. S San Antonio de las Alazanas, 8900 ft., January 11, 1954; and 31654 from Sierra Guadalupe (=10 mi. S, 5 mi. W General Cepeda), 7000 ft., April 13, 1953.

The Mexican Jay is common in Coahuila. This species occupies various habitats in the State and has been collected at stations ranging from 2100 to 9950 feet.

Miller (1955a:167) stated that the Mexican Jay was the most abundant species of bird in the Sierra del Carmen. Burleigh and Lowery (1942:196) remarked that specimens of *A. u. couchii* were obtained "in the area surrounding the summit of Diamante Pass." At no time did Burleigh and Lowery (*loc. cit.*) see this species below 6500 feet. The Mexican Jay is restricted to the higher altitudes in southern Coahuila but is more widespread in northern Coahuila. Dickerman saw Mexican Jays in the Sierra del Pino on May 12, 1954; 16 mi. E and 18 mi. N Ocampo on May 7, 1954; in the Sierra de la Madera on December 13, 1953; and 20 mi. S Ocampo on April 4, 1954.

Ridgway (1904:340) remarked that *Aphelocoma sieberii potosina* Nelson [=*A. u. couchii* (Baird)] has been recorded in southern Coahuila, at Carneros. Miller, Friedmann, Griscom, and Moore (1957:124) recorded *A. u. couchii* from Carneros, Nuevo Leon. I suspect (Ridgway, 1904:340 and Goldman, 1951: map opposite p. 34) that the locality given by Miller, Friedmann, Griscom, and Moore (*op. cit.*) should be Carneros, Coahuila, rather than Carneros, Nuevo Leon.

Specimens (31051, 29425-29428, 31635-31636, 32081-32082, 31607, and 31654) of the Mexican Jay increase in size from northern Coahuila to southern Coahuila. The average length of the wings of Nos. 29425-29426 from Club Sierra del Carmen is 152 mm. whereas the average length of the wings of No. 32081 from 7 mi. S and 4 mi. E Bella Union, of No. 31607 from 13 mi. E and 3 mi. S San Antonio de las Alazanas, and of No. 31654 from Sierra Guadalupe is 164 mm.

Miller (1955a:169) indicated that the Mexican Jay breeds in the Sierra del Carmen. Burleigh and Lowery (1942:196) remarked that the species breeds at Diamante Pass. The sizes of the testes (12, 11 mm.) of Nos. 29425-29426, the size of the ovum (8 mm.) of No. 29428, and the short tail (126 mm.) of the immature female from 4 mi. N and 21 mi. W Cuatro Cienegas are also evidence of breeding by this species in the State.

****Corvus corax sinuatus*** Wagler.--The Common Raven seems to occur in low density in Coahuila. Miller (1955a:168) saw ravens in the pine-oak and cliff areas of the Sierra del Carmen, and took a breeding female at the head of Corte Madera Canyon, 7500 feet. Burleigh and Lowery (1942:196) found this species to be a bird of the higher slopes of the mountains although not uncommon in the broad open valley south of Diamante Pass. Sutton and Burleigh (1939a:35) took a male at Santa Rosa.

****Corvus cryptoleucus*** Couch.--*Specimens examined:* total 4: 32080 from 2 mi. W Jimenez, 850 ft., June 20, 1952; 35404 (skeleton only) from 4 mi. N San Isidro (=16 mi. N Ocampo), May 10, 1954; and 31474-31475 from R. de Almendarez (=53 mi. NW Ocampo), May 11, 1954.

The White-necked Raven occurs throughout Coahuila. Sutton and Burleigh (1939a:36) observed this species "in the vicinity of Saltillo," but not farther west. Burleigh and Lowery (1942:197) noted *C. cryptoleucus* frequently "on the arid plateau around Saltillo" and obtained a specimen "in the high fertile valley south of Diamante Pass." Burleigh and Lowery (*loc. cit.*) remarked that the White-necked Raven avoids the mountain slopes; 7000 feet was the uppermost limit of occurrence. The sizes of the testes (32080, 20 mm.; 31474: left, 12x20 mm., right, 10x16 mm.; and 31475: left, 10x16 mm., right, 10x14 mm.) of the specimens that I have examined and the dates (May 11, June 20) on which they were collected indicate breeding by the White-necked Raven in Coahuila.

****Parus sclateri eidos*** (Peters).--*Specimens examined:* total 4: 32083 from 7 mi. S, 4 mi. E Bella Union, 7200 ft., June 25, 1952; 31609 from 2 mi. E Mesa de Tablas, 9000 ft., January 15, 1954; 31656 and 31655 from the north slope of Sierra Guadalupe (=11 mi. S, 7 mi. W General Cepeda), 7800 ft., April 20, 1953.

The Mexican Chickadee is common in southeastern Coahuila. Miller, Friedmann, Griscom, and Moore (1957:133) stated that *P. s. eidos* and *P. s. sclateri* intergrade in southern Coahuila. The specimens that I have examined also show signs of intergradation, but are closer to *P. s. eidos*.

Sutton and Burleigh (1939a:36) reported taking specimens of *P. s. eidos* at Diamante Pass where they saw several individuals. Burleigh and Lowery (1942:197) also observed these chickadees "in the pine woods about the summit of Diamante Pass...." The size of the testes (6x5 mm.) of No. 31656 and the fact that No. 31655 was incubating eggs at the time it was obtained are evidence of breeding by this chickadee in the State.

****Parus sclateri sclateri** Kleinschmidt.--Miller, Friedmann, Griscom, and Moore (1957:133) stated that a specimen (or specimens?) of *P. s. sclateri* which showed evidence of intergradation with *P. s. eidos* was obtained at El Diamante.

****Parus atricristatus dysleptus** Van Tyne.--*Specimen examined:* one, 31054 from 37 mi. S, 21 mi. E Boquillas, 4100 ft., March 13, 1952.

Two subspecies of the Black-crested Titmouse are present in Coahuila. *P. a. dysleptus* occurs in northwestern Coahuila. Miller (1955a:168) stated that the Black-crested Titmouse, identified as *dysleptus,* was the only representative of the genus *Parus* in the Sierra del Carmen. The weak extension of black onto the nape in No. 31054 suggests intergradation between *P. a. dysleptus* and *P. a. atricristatus*; the latter lacks the black nape of *dysleptus*.

***Parus atricristatus atricristatus** Cassin.--*Specimens examined:* total 4: 32084 from 12 mi. N, 12 mi. W Jimenez, 850 ft., June 19, 1952; sex ? 32085 from 2 mi. W Jimenez, 850 ft., June 20, 1952; 31055 (skeleton only) from Fortin (=33 mi. N, 1 mi. E San Geronimo), 3300 ft., on March 29, 1952; and 35399 (skeleton only) from 3.5 mi. W, 22 mi. S Ocampo, December 15, 1953, weight, 16 gms.

Typical *P. a. atricristatus* occurs in Coahuila in the southeastern sector (Dixon, 1955:184). Black-crested Titmice intermediate between *atricristatus* and *dysleptus* were listed from Cuidad Muzquiz and Sabinas by Dixon (*loc. cit.*:189), as *dysleptus* but were shown on his map (*loc. cit.*:184) as *atricristatus*. Our Nos. 32084 and 32085 (wing, 71, 71, tail, 63, 64 mm.) are small and fall in the upper range of size for *atricristatus* to which the specimens are here referred.

***Auriparus flaviceps ornatus** (Lawrence).--*Specimen examined:* one, 31056, from Sierra de la Encantada (=38 mi. S, 23 mi. E Boquillas), 4400 ft., March 15, 1952.

The Verdin occurs up to about 5000 feet in Coahuila. Miller (1955a:168) reported that "this desert species followed the catclaw scrub up the washes to about 4800 feet, the limit of such habitat" in the Sierra del Carmen and also gave evidence of breeding by the Verdin in Coahuila. Amadon and Phillips (1947:578) reported a nest of *Auriparus flaviceps* at Las Delicias. Burleigh and Lowery (1942:197) obtained a single specimen "in an arroyo a few miles east of Saltillo" and a single specimen "about twenty miles west" of Saltillo. Ridgway (1904:421) reported *A. f. ornatus* [he referred to it as *A. f. flaviceps*] at Monclova. Hellmayr (1934:88) listed *A. f. ornatus* from Jaral. Findley saw a Verdin 2 mi. S and 3 mi. E San Juan de Sabinas on June 22, 1952.

The size (wing, 50 mm.; tail, 43 mm.) of No. 31056 is small for typical representatives of *A. f. ornatus*. The yellow of the head of No. 31056 is darker than that of the other subspecies of the Verdin, and I have accordingly allocated the specimen to *A. f. ornatus*.

***Psaltriparus melanotis lloydi** Sennett.--*Specimens examined:* total 5: 31058 and 31057 from 37 mi. S, 21 mi. E Boquillas, 4100 ft., March 13, 1952; 31060 and 31059 from Sierra de la Encantada (=38 mi. S, 23 mi. E Boquillas), 4400 ft., March 15, 1952; and 35407 (skeleton only) from Sierra del Pino (=5 mi. W, 3 mi. S Acebuches), 6200 ft., May 14, 1954.

The Black-eared Bushtit is common in Coahuila. Typical representatives of *P. m. lloydi* in the northern part of the State range from 4100 feet to 8000 feet. Miller (1955a:168-169) reported *P. m. lloydi* breeding in the Sierra del Carmen. Marsh and Stevenson (1938:287) obtained a male in the Sierra del Carmen at El Jardin. The size of the testes (3.5 mm.) of No. 31058 suggests breeding by the Black-eared Bushtit 37 mi. S and 21 mi. E Boquillas.

****Psaltriparus melanotis iulus** Jouy.--*Specimens examined:* total 2: 31659 from the north slope of Sierra Guadalupe (=10 mi. S, 7 mi. W General Cepeda), 7000 ft., April 20, 1953; and 31658 from the north slope of Sierra Guadalupe (=10 mi. S, 5 mi. W General Cepeda), 6700 ft., April 15, 1953.

Typical representatives of this Black-eared Bushtit are present in the southeastern sector of Coahuila. The backs of Nos. 31658-31659 differ slightly in color from the backs of typical representatives of *lloydi*. I suspect that the specimens from the Sierra Guadalupe are intergrades between *lloydi* and *iulus*.

Burleigh and Lowery (1942:197) took two males and two females of *Psaltriparus melanotis iulus* (they called their specimens *Psaltriparus minimus iulus*) at Saltillo and stated that their specimens tended to approach *lloydi* rather than being typical *iulus*. Sutton and Burleigh (1939a:36) recorded *P. m. iulus* only at Diamante Pass where they took two specimens.

***Sitta carolinensis nelsoni** Mearns.--Miller (1955a:169) reported that this subspecies of the White-breasted Nuthatch breeds and was common in the oaks and open conifers from 6500 to 8000 feet in the Sierra del Carmen and stated also that the populations of the White-breasted Nuthatch in "the Chisos Mountains [of Texas] and the Sierra del Carmen seem best regarded as a stage in the cline of which *nelsoni* and *mexicana* are end points, although falling closer to *nelsoni*."

Sitta carolinensis mexicana* Nelson and Palmer.--*Specimens examined:* total 2: 31669 and 31670 from the Canon d. Meco, Sierra Guadalupe (=10 mi. S General Cepeda), 6500 ft., April 23, 1953, weights, 17, 18 gms.

Typical populations of this White-breasted Nuthatch occur in southern Coahuila. Ridgway (1904:449) listed *S. c. mexicana* from Sierra Guadalupe in southern Coahuila. The underparts of our specimens are darker than in *nelsoni*, and their bills (culmen, 15, 13 mm.) are shorter than the average (, 19.8 mm.; , 18.6 mm. [Ridgway, 1904:447]) in *nelsoni*. The large size of the testes (5x3 mm.) of No. 31669 suggests breeding by *S. c. mexicana* in Sierra Guadalupe.

*****Sitta pygmaea melanotis*** van Rossem.--In Coahuila the Pigmy Nuthatch seems to be locally common. Miller (1955a:169) reported it so between 7500 and 8000 feet in pine-oak on the mesa tops and in the heads of canyons of the Sierra del Carmen and noted that it breeds there. Hardy saw the Pigmy Nuthatch 13 mi. E San Antonio de las Alazanas on July 6, 1955.

Certhia familiaris americana Bonaparte.--*Specimens examined:* total 2: 31612 from the base of Don Martin Dam, November 27, 1953, skull partially unossified; and 31587 from 20 mi. S Ocampo, 6500 ft., April 5, 1954, weight, 7 gms.

This subspecies of the Brown Creeper can be considered a sparse winter visitant to Coahuila. Van Hoose (1955:302) reported that Nos. 31612 and 31587 constitute the southernmost records of *C. f. americana* and represent the first records of occurrence of *americana* in Mexico.

Certhia familiaris montana Ridgway.--Miller (1955a:169) reported this subspecies of the Brown Creeper, which he assumed to be a winter visitant or a migrant, in the Sierra del Carmen. He (*loc. cit.*) remarked also that the higher conifers would seem to constitute favorable habitat for nesting by the Brown Creeper, but did not find any evidence of a breeding population of creepers in the Sierra del Carmen. Miller, Friedmann, Griscom, and Moore (1957:143) reported *C. f. montana* from San Lazaro Mountain on November 9.

*****Certhia familiaris albescens*** Berlepsch.--*Specimens examined:* total 3: sex ? 32805 from 13 mi. E San Antonio de las Alazanas, July 7, 1955; 31610-31611 from 3 mi. S, 13 mi. E San Antonio de las Alazanas, 8900 ft., January 12, 1954.

Miller, Friedmann, Griscom, and Moore (1957:143) reported *C. f. albescens* from "southern Coahuila." Nos. 31610-31611 and 32805 represent the only other records of this subspecies from the State. The date (July 7) on which No. 32805 was obtained suggests that this bird was a resident 13 mi. E San Antonio de las Alazanas.

Troglodytes aedon parkmanii Audubon.--*Specimen examined:* one, sex ? 29556, from 1.5 mi. N Parras, 5500 ft., November 10, 1949, weight, 9.8 gms.

Burleigh and Lowery (1942:197) recorded the House Wren "in small numbers about Saltillo where occasional birds, presumably migrants, were noted in thickets or stretches of underbrush fringing cultivated fields." They obtained a single male "on the outskirts of Saltillo." Hellmayr (1934:218) listed *T. a. parkmanii* from Sabinas.

****Troglodytes brunneicollis cahooni*** Brewster.--Typical representatives of this subspecies of the Brown-throated Wren occur in northern Coahuila. In the Sierra del Carmen, Miller (1955a:170) found *T. b. cahooni* that in no way suggested *compositus* of the Sierra Madre Oriental. Burleigh and Lowery (1942:198) recorded a *cahooni*-like specimen from Diamante Pass in southern Coahuila.

*****Troglodytes brunneicollis compositus*** Griscom.--*Specimen examined:* one, 32819, from 13 mi. E San Antonio de las Alazanas, July 6, 1955.

The subspecies *cahooni* and *compositus* of the Brown-throated Wren seem to intergrade in the southern part of the State. Although No. 32819 represents the subspecies *compositus*, the somewhat whitish abdomen and the fairly large spots of the lesser wing coverts suggest some relationship with *cahooni*. In addition to the present record, Miller, Friedmann, Griscom, and Moore (1957:163) recorded *T. b. compositus* from southern Coahuila at Sierra Guadalupe. The record of *T. b. cahooni* from Sierra Guadalupe (Ridgway, 1904:588) I suspect probably represents *T. b. compositus* or an intergrade between *compositus* and *cahooni*. The date (July 6) on which No. 32819 was obtained 13 mi. E San Antonio de las Alazanas suggests that this bird was resident there.

****Thryomanes bewickii eremophilus*** Oberholser.--*Specimens examined:* total 3: 32088 from 2 mi. W Jimenez, 850 ft., June 20, 1952; 31061 from 4 mi. W Hacienda La Mariposa, 2300 ft., March 24, 1952, weight, 10.8 gms.; and 31660 from the north foot of Sierra Guadalupe (=10 mi. S, 5 mi. W General Cepeda), 6500 ft., April 21, 1953, weight, 13 gms.

Bewick's Wren occurs commonly in Coahuila. Miller, Friedmann, Griscom, and Moore (1957:160) reported that, in Coahuila, *T. b. eremophilus* "intergrades in the eastern and southern sections with *T. b. cryptus* and *T. b. murinus*, respectively." The slightly darker coloration of No. 31660, suggesting a resemblance to *T. b. murinus*, is the only evidence of intergradation of *T. b. murinus* and *eremophilus* that I have found.

Miller (1955a:170) stated that *T. b. eremophilus* was "common in the piedmont area on yucca-dotted slopes and along the lower canyon walls in growth of pinon, yucca, and cactus" in the Sierra del Carmen, and reported breeding there. Burleigh and Lowery (1942:198) remarked that *T. b. eremophilus* "proved without question to be the most widely distributed and abundant wren" in the Saltillo region. The series that Burleigh and Lowery (*loc. cit.*) assembled "proved to be uniform and clearly referable to" *T. b. eremophilus*. Sutton and Burleigh (1939a:36) noted *Thryomanes bewickii* at San Pedro on January 29 and 30. Hardy saw a male *T. b. eremophilus* at Parras on July 4, 1955. Ridgway (1904:557) listed *T. b. eremophilus* from Saltillo in April and as breeding at Sabinas.

The sizes of the testes (8 mm.; 6x4 mm.) of Nos. 32088 and 31660, respectively, suggest breeding 2 mi. W Jimenez and Sierra Guadalupe.

Thryomanes bewickii cryptus Oberholser.--Miller, Friedmann, Griscom, and Moore (1957:161) recorded *T. b. cryptus* from Saltillo.

Thryothorus ludovicianus berlandieri* Baird.--*Specimens examined:* total 3: 32086 from 12 mi. N, 12 mi. W Jimenez, 850 ft., June 19, 1952; sex ? 32087 from 2 mi. W Jimenez, June 20, 1952; and 31063 from 8 mi. N, 4 mi. W Muzquiz, 1800 ft., April 1, 1952, weight, 18.3 gms.

One subspecies of the Carolina Wren, *berlandieri*, occurs in Coahuila in the northeastern section of the State. Ridgway (1904:547) recorded *T. l. berlandieri* from Sabinas. The fact that No. 32086 was a juvenile suggests that the Carolina Wren breeds 12 mi. N and 12 mi. W Jimenez.

Campylorhynchus brunneicapillum couesi* Sharpe.--*Specimens examined:* total 6: 29429 from Canon del Cochino (=16 mi. N, 21 mi. E Piedra Blanca), 3200 ft., April 6, 1950; 31064 from 7 mi. S, 2 mi. E Boquillas, 800 ft., February 29, 1952, weight, 38.1 gms.; 31066 (skeleton only) from 10 mi. S, 5 mi. E Boquillas, 1500 ft., March 5, 1952; 31637 from La Gacha (=La Concha), December 1, 1953, weight, 40 gms.; 31638 from 18 mi. S Ocampo, December 16, 1953; and sex ? 29557 from 7 mi. S, 1 mi. E Gomez Farias, 6500 ft., November 18, 1949, weight, 41 gms.

This subspecies of the Cactus Wren occurs throughout Coahuila except in the extreme southeastern section of the State, where the subspecies *C. b. guttatus* occurs. Miller (1955a:169) found *C. b. couesi* breeding and occupying the open swales and mesas at the base of the mountains of the Sierra del Carmen. Burleigh and Lowery (1942:198) found *C. brunneicapillum* "to be rather scarce and decidedly local in its distribution" and observed an occasional bird "in the open desert country west of Saltillo." Miller, Friedmann, Griscom, and Moore (1957:151) recorded *C. b. couesi* "south to [the] vicinity of Monclova."

No. 29557 does not have any characters of *C. b. guttatus*; its under tail coverts and flanks have roundish black spots, rather than black bars as in *C. b. guttatus*.

**Campylorhynchus brunneicapillum guttatus* (Gould).--This subspecies of Cactus Wren seems to occupy the extreme southeastern section of Coahuila. Miller, Friedmann, Griscom, and Moore (1957:152) recorded *C. b. guttatus* from Hipolito.

Telmatodytes palustris plesius (Oberholser).--Miller, Friedmann, Griscom, and Moore (1957:148) reported this subspecies of the Long-billed Marsh Wren from 8 mi. S Cuatro Cienegas.

Catherpes mexicanus albifrons (Giraud).--Miller (1955a:170) found this subspecies of Canon Wren "in shaded rocky canyons and on larger cliff slopes at the base of the mountains from 4700 to 5300 feet" in the Sierra del Carmen where it nested. Burleigh and Lowery (1942:198) noted that the Canon Wren was "decidedly uncommon" at Saltillo and obtained a male at the Chorro de Agua on April 19. Ridgway (1904:657) listed *C. m. albifrons* from Patos.

Salpinctes obsoletus obsoletus (Say).--*Specimens examined:* total 4: 31067 from 1 mi. N Boquillas, 700 ft., March 6, 1952, weight, 16.1 gms.; 31068 from 7 mi. S, 2 mi. E Boquillas, 800 ft., March 1, 1952, weight, 18.2 gms.; sex ? 29558 from 12 mi. N, 10 mi. E Parras, 5000 ft., November 11, 1949, weight, 16.9 gms.; and 32089 from 7 mi. S, 4 mi. E Bella Union 7200 ft., June 24, 1952.

The Rock Wren is common in Coahuila. Miller (1955a:170) found *S. o. obsoletus* "only in the rocky piedmont and on lower bare canyon faces" and stated that Marsh took a bird in fresh fall plumage on September 6 at El Jardin. Sutton and Burleigh (1939a:37) found the Rock Wren "near San Pedro." Burleigh and Lowery (1942:198) wrote that the subspecies *obsoletus* was "characteristically a bird of the arroyos of the arid plateau about Saltillo, where it was fairly common...." The large size of the testes (5x3 mm.) of No. 32089 and the date (June 24) on which it was obtained suggest breeding by the Rock Wren 7 mi. S and 4 mi. E Bella Union.

Mimus polyglottos leucopterus (Vigors).--*Specimens examined:* total 5: 31070 from 10 mi. S, 5 mi. E Boquillas, 1500 ft., March 5, 1952, weight, 55.1 gms.; 32094-32095 from 2 mi. W Jimenez, 850 ft., June 20, 1952; 32096 from 5 mi. N, 19 mi. W Cuatro Cienegas, 3250 ft., July 5, 1952; and 33186 (skeleton only) from Parras, July 5, 1955, testes, 6x3 mm.

The Mockingbird is sparsely distributed throughout Coahuila. Miller (1955a:170) found *M. p. leucopterus* in the mesquite and catclaw at the base of the mountains in the Sierra del Carmen. Sutton and Burleigh (1939a:37) reported *M. p. leucopterus* from Diamante Pass. Amadon and Phillips (1947:578) found a young Mockingbird out of the nest begging for food from an adult on August 18 at Las Delicias. Burleigh and Lowery (1942:199) found the Mockingbird on the arid plateau "about Saltillo." Hellmayr (1934:308) listed *M. p. leucopterus* from Jaral. Findley saw Mockingbirds 2 mi. S and 3 mi. E San Juan de Sabinas on June 22, 1952. Dickerman saw Mockingbirds in the Sierra del Pino on May 12, 1954, and 8 mi. E and 2 mi. S Americanos on May 18, 1954. The sizes of the testes (8, 7 mm.) of Nos. 32094 and 32096, respectively, suggest breeding 2 mi. W Jimenez and 5 mi. N and 19 mi. W Cuatro Cienegas, as does No. 32095, a juvenile.

****Toxostoma longirostre sennetti*** (Ridgway).--*Specimen examined:* one, 32090, from 2 mi. S, 3 mi. E San Juan de Sabinas, June 22, 1952.

In Coahuila the Long-billed Thrasher seems to be uncommon. It has been recorded in Coahuila as far west as San Juan de Sabinas. Ridgway (1907:192) recorded *T. l. sennetti* from Sabinas, the only other record of the Long-billed Thrasher, to my knowledge, from the State. The large size of the testes (11x6 mm.) of No. 32090 and the date (June 22) on which it was obtained suggest that *T. l. sennetti* breeds 2 mi. S and 3 mi. E San Juan de Sabinas.

****Toxostoma curvirostre celsum*** Moore.--*Specimens examined:* total 2: 31071 from 7 mi. S, 2 mi. E Boquillas, March 1, 1952, wing, 111 mm., tail, 114 mm., weight, 97.2 gms.; and 31072 (skeleton only) from 10 mi. S, 5 mi. E Boquillas, 1500 ft., March 5, 1952.

This subspecies of the Curve-billed Thrasher occurs in northwestern Coahuila. Specimens of *T. c. celsum* and *oberholseri* from Coahuila are too few to show clearly the distribution and intergradation in Coahuila.

No. 31071 is referred to *T. c. celsum* because of large size; the spots on its upper abdomen, which are large and pronounced, suggest a relationship with *T. c. oberholseri*. Miller (1955a:170) remarked that *T. c. celsum* was a scarce resident of the desert scrub at the mouth of Boquillas Canyon of the Sierra del Carmen.

****Toxostoma curvirostre oberholseri*** Law.--*Specimens examined:* total 5: 35405 (skeleton only) from 4 mi. N San Isidro, May 11, 1954; 32091 from 5 mi. N, 19 mi. W Cuatro Cienegas, 3250 ft., July 5, 1952; 32833 from Parras, July 4, 1955, weight, 76.5 gms.; 32092 from 7 mi. S, 4 mi. E Bella

Union, 7200 ft., June 25, 1952; and 31614 from 16 mi. W San Antonio de las Alazanas, 6500 ft., January 7, 1954, weight, 90 gms.

This subspecies of the Curve-billed Thrasher occurs in eastern and southern Coahuila. Amadon and Phillips (1947:578) took a Curve-billed Thrasher twenty miles west of Saltillo that had an enlarged ovary and a brood patch still somewhat evident on August 27. Burleigh and Lowery (1942:199) stated that *T. c. oberholseri* "was rather widely and commonly distributed, being noted from the area about the summit of Diamante Pass at 7,800 feet down to the desert country about Saltillo." Miller, Friedmann, Griscom, and Moore (1957:177) recorded *T. c. oberholseri* from Sabinas, from 8 mi. S Cuatro Cienegas, and from El Diamante. Hellmayr (1934:298) also recorded *T. c. oberholseri* from Sabinas, as did Ridgway (1907:199) under the name *T. c. curvirostre* before the subspecies *oberholseri* was named. The female from 4 mi. N San Isidro had an egg in its oviduct. The immature male (32833), the large size of ovum (8 mm.) of No. 32092, and the presence of a brood patch on No. 32091 also are evidences of breeding by the Curve-billed Thrasher in Coahuila.

*****Toxostoma dorsale dorsale*** Henry.--The Crissal Thrasher is uncommon in Coahuila. The subspecies *dorsale* occurs in northern Coahuila. Miller (1955a:170-171) found the subspecies *dorsale,* at about 4700 feet, only in the mesquite, desert willow, and walnut scrub along the wash of Boquillas Canyon of the Sierra del Carmen and remarked also that the bird nested there.

*****Toxostoma dorsale dumosum*** Moore.--*Specimen examined:* one, sex ? 29559, from 8 mi. N La Ventura, 6000 ft., November 17, 1949, weight, 57.0 gms.

The subspecies *dumosum* of the Crissal Thrasher in Coahuila has been reported only from the southeastern section of the State. Burleigh and Lowery (1942:199-200) found *T. d. dumosum* "not uncommon in the lower foothills outside of Saltillo as well as on the summit of Diamante Pass." The specimen of *T. d. dorsale* from Diamante Pass reported by Sutton and Burleigh (1939a:37) is closer, according to Burleigh and Lowery (1942:199), to *T. d. dumosum*. No. 29559 is darker above and below than typical specimens of *T. d. dorsale* and represents *T. d. dumosum.*

Oreoscoptes montanus (Townsend).--*Specimen examined:* one, sex ? 30237, from 1 mi. SW San Pedro de las Colonias, 3700 ft., February 8, 1951.

The Sage Thrasher seems to be a winter visitant to Coahuila. Miller, Friedmann, Griscom, and Moore (1957:173) recorded the species in November from 8 mi. S Cuatro Cienegas.

****Turdus migratorius propinquus*** Ridgway.--*Specimen examined:* one, 31073 (skeleton only) from 4 mi. W Hacienda La Mariposa, 2300 ft., March 24, 1952.

Burleigh and Lowery (1942:200) stated that "the Robin apparently breeds rather sparingly on the higher ridges" in southeastern Coahuila. They collected a pair "in the open pine woods just below the summit of Diamante Pass" on April 15 and noted another at the Chorro del Agua on April 19.

*****Ridgwayia pinicola*** (Sclater).--*Specimen examined:* one, 31619, from 5 mi. W, 22 mi. S Ocampo, 6000 ft., December 15, 1953, weight, 88 gms.

The Aztec Thrush is rare in Coahuila. Van Hoose (1955:302) remarked that No. 31619, the skull of which was incompletely ossified, "represents the northernmost record for this species, which was previously unknown in Coahuila."

Hylocichla guttata guttata (Pallas).--*Specimens examined:* total 2: 31074-31075 from 4 mi. W Hacienda La Mariposa, 2300 ft., March 25 and 26, 1952, weights, 25, 21 gms.

The Hermit Thrush is a common migrant or winter visitant in Coahuila. *H. g. guttata* has been reported from northern Coahuila. Miller (1955a:171) observed *H. g. guttata* (and *H. g. auduboni*) in the Douglas fir and pine-oak belts and in the lower levels in the oaks at the foot of the Sierra del Carmen. Burleigh and Lowery (1942:200) noted *H. g. guttata* "in small numbers in the open woods surrounding the summit of Diamante Pass, and at infrequent intervals in the arroyos on the arid plateau near Saltillo."

Hylocichla guttata sequoiensis (Belding).--Ridgway (1907:45) recorded this subspecies of Hermit Thrush from Sierra Guadalupe in April. However, Miller, Friedmann, Griscom, and Moore (1957:188) suggest that the material on which this identification was based needs redetermination.

Hylocichla guttata auduboni (Baird).--*Specimens examined:* total 3: 31488 from Sierra del Pino (=5 mi. S, 3 mi. W Acebuches), May 12, 1954, weight, 25 gms.; 31076 (skeleton only) from Fortin (=33 mi. N, 1 mi. E San Geronimo), 3300 ft., March 28, 1952; and 31077 (skeleton only) from 26 mi. W Santa Teresa, 7050 ft., April 4, 1952.

Miller (1955a:171) found this subspecies of Hermit Thrush wintering with *H. g. guttata* in the Sierra del Carmen. Hellmayr (1934:456) listed *H. g. auduboni* from Sabinas.

****Sialia sialis fulva** Brewster.--Hellmayr (1934:479) listed this subspecies of the Eastern Bluebird from Sabinas.

***Sialia mexicana mexicana** Swainson.--Miller, Friedmann, Griscom, and Moore (1957:199) remarked that the subspecies *mexicana* of the Western Bluebird breeds in the southern mountains of Coahuila where, at El Diamante, on July 7, a specimen (or specimens ?) in breeding condition was obtained. Burleigh and Lowery (1942:200) found *S. m. mexicana* "well distributed in the open woods about" Diamante Pass, but at no time below an elevation of approximately 6500 feet. Sutton and Burleigh (1939a:38) also saw "several brown-backed" Western Bluebirds at Diamante Pass on March 6. Ridgway (1907:150) recorded *S. m. mexicana* from Saltillo, Carneros, and Sierra Guadalupe.

Sialia currucoides (Bechstein).--*Specimen examined:* one, 31078, from Sierra de la Encantada (=38 mi. S, 23 mi. E Boquillas), 4400 ft., March 14, 1952, weight, 23.7 gms.

The Mountain Bluebird is a winter visitant to Coahuila. Miller, Friedmann, Griscom, and Moore (1957:200) recorded *Sialia currucoides* from Hipolito on February 24.

****Myadestes townsendi townsendi** (Audubon).--*Specimen examined:* one, 31079, from 26 mi. W Santa Teresa, 7050 ft., April 4, 1952.

Miller (1955a:171) detected Townsend's Solitaire in clumps of large pines in two different locations at 7000 and 7500 feet on April 4, 6, and 8 in the Sierra del Carmen. He did not find a breeding population of *M. t. townsendi,* but did note favorable habitat for breeding. Ridgway (1907:164) recorded *M. townsendi* from the Sierra Guadalupe on April 21. Dickerman saw a Townsend's Solitaire in the Sierra de la Madera on December 13, 1953.

The underparts of No. 31079 are not uniformly dark, being paler on the chin, throat, and abdomen than elsewhere as is true of typical representatives of *M. t. townsendi.*

****Polioptila caerulea caerulea** (Linnaeus).--*Specimen examined:* one, 32097, from 2 mi. S, 11 mi. E Nava, June 15, 1952.

No. 32097 is the first record of occurrence of this subspecies of Blue-gray Gnatcatcher in the State. The white of the underparts of No. 32097 is less grayish than the underparts of typical representatives of *P. c. amoenissima,* and the black at the base of the inner webs of the outermost rectrix does not extend beyond the tip of the under tail coverts.

Representatives of *P. c. amoenissima* have black at the base of the inner web of the outermost rectrix more extended, usually showing beyond the tip of the under tail coverts. The size of the testes (3x2 mm.) of No. 32097 does not suggest breeding, but the date (June 15) indicates that it was a resident.

Polioptila caerulea amoenissima* Grinnell.--*Specimen examined:* one, 31080, from 7 mi. S, 2 mi. E Boquillas, 800 ft., February 29, 1952, weight, 5.4 gms.

This subspecies of the Blue-gray Gnatcatcher breeds in Coahuila and occurs throughout all of the State except the northeastern section. Miller (1955a:171) remarked that "this gnatcatcher was apparently established on summer territories in the oaks and walnuts of the wash of Boquillas Canyon at the foot of the mountains" of the Sierra del Carmen. Burleigh and Lowery (1942:200) noted "this species only in the open woods at the summit of Diamante Pass, where, however, it was not uncommon." Miller, Friedmann, Griscom, and Moore (1957:201) recorded *P. c. amoenissima* breeding at El Diamante on July 8.

Polioptila melanura melanura Lawrence.--*Specimen examined:* one, 31081, from 7 mi. S, 2 mi. E Boquillas, 800 ft., March 1, 1952, weight, 5.6 gms.

In Coahuila this subspecies of the Black-tailed Gnatcatcher has been recorded from several localities. Burleigh and Lowery (1942:200) found that it "was limited in its distribution to the lower altitudes and was noted only in the open desert country west of Saltillo." Sutton and Burleigh (1939a:38) noted it "several times near San Pedro" where on January 29 one female was taken. Miller, Friedmann, Griscom, and Moore (1957:205) recorded *P. m. melanura* at Hipolito on June 30 to July 2 in breeding condition.

Regulus satrapa satrapa Lichtenstein.--Miller (1955a:171) found a small wintering flock of Golden-crowned Kinglets in the Sierra del Carmen; this is the only record of the species in Coahuila.

Regulus calendula calendula (Linnaeus).--*Specimens examined:* total 5: 31085 from the Rio Grande (=17 mi. S Dryden, Terrell Co., Texas, in Coahuila), 600 ft., March 19, 1952, weight, 6.3 gms.; 31082-31083 from 1 mi. N Boquillas, 700 ft., March 8, 1952, weight, 7.2, 6.5 gms.; 31084 from Sierra de la Encantada (=38 mi. S, 23 mi. E Boquillas), 4400 ft., March 15, 1954, weight, 5.4 gms.; and 31661 from the north slope of Sierra Guadalupe (=11 mi. S, 7 mi. W General Cepeda), 7800 ft., April 20, 1953, weight, 5 gms.

In Coahuila this subspecies of the Ruby-crowned Kinglet is a common migrant. Miller (1955a:171) found it (and *R. c. cineraceus*) "common in the conifers and oaks of the upper levels of the mountains [Sierra del Carmen], at 6500 to 7000 feet, as winter visitants or migrants." Burleigh and Lowery (1942:201) took specimens of *R. c. calendula* at Diamante Pass on April 15, at the Chorro del Agua on April 19, and at 20 mi. W Saltillo on April 22. Sutton and Burleigh (1939a:38) noted the Ruby-crowned Kinglet "in the arid parts of southern Coahuila." Hellmayr (1934:513) listed *R. c. calendula* from Sabinas. Dickerman saw Ruby-crowned Kinglets in the Sierra de la Madera on December 13, 1953, 20 mi. S Ocampo on April 4, 1954, and 3 mi. S and 13 mi. E San Antonio de las Alazanas on January 12, 1954.

Regulus calendula cineraceus Grinnell.--Miller (1955a:171) found *R. c. cineraceus* common in the Sierra del Carmen; on April 3, 5, and 10 the birds were "abundant, as though a wave of migrants were passing through."

Anthus spinoletta rubescens (Tunstall).--*Specimens examined:* Total 3: 31086-31087 and sex ? 31088 from 1 mi. N Boquillas, 700 ft., March 6, 7, and 8, 1952, weights, 19.3, 19.9, and 16.6 gms.

This subspecies of the Water Pipit is an uncommon winter visitant or migrant in Coahuila. Burleigh and Lowery (1942:201) found "a flock of ten birds ... on the outskirts of Saltillo" on April 18. Miller, Friedmann, Griscom, and Moore (1957:210) recorded *A. s. rubescens* from Cuatro Cienegas in November and from Hipolito in February.

Anthus spinoletta pacificus Todd.--Miller, Friedmann, Griscom, and Moore (1957:209) recorded this Water Pipit from Cuatro Cienegas in February and from Hipolito in November.

Bombycilla cedrorum Vieillot.--The Cedar Waxwing is an uncommon winter visitant to Coahuila. Miller (1955a:171) recorded a flock in the Sierra del Carmen on April 5, and another flock on April 21. Burleigh and Lowery (1942:201) saw two small flocks on April 15 "in the open woods just below the summit of Diamante Pass."

****Phainopepla nitens nitens*** Swainson.--*Specimen examined:* one, 31674, from the west foot of Pico de Jimulco, 5000 ft., April 3, 1953, weight, 35 gms.

The Phainopepla occurs throughout most of Coahuila. Sutton and Burleigh (1939a:39) reported *P. n. nitens* from Diamante Pass on March 6. On April 15 and 17 Burleigh and Lowery (1942:201) saw scattered pairs of the Phainopepla "only in the open woods surrounding Diamante Pass." Miller (1955a:171) noted *P. n. nitens* "on April 20 and 28 in large clumps of mesquite near Piedra Blanca, at about 4500 feet, on the foothills" of the Sierra del Carmen. Dickerman saw a Phainopepla 20 mi. S Ocampo on April 4, 1954. Baird (1858:320) listed a male *P. n. nitens* from Coahuila, Mexico. Hellmayr (1935:107) remarked that *Phainopepla nitens* was listed from Coahuila by "Salvin and Godman, Biol. Centr.--Amer., Aves, 1, p. 220, 1883...." Miller, Friedmann, Griscom, and Moore (1957:213) stated that *P. n. nitens* breeds at El Diamante. The long wing (100 mm.) and long tail (96 mm.) of No. 31674 is typical for *P. n. nitens.*

Phainopepla nitens lepida Van Tyne.--Sutton and Burleigh (1939a:39) recorded *P. n. lepida* from Diamante Pass on March 6; Miller, Friedmann, Griscom, and Moore (1957:213) suggest that this individual was a vagrant.

Lanius ludovicianus migrans Palmer.--Burleigh and Lowery (1942:202) obtained this subspecies of Loggerhead Shrike "on the outskirts of Saltillo on April 20." The specimen shows evidence of intergradation with *excubitorides.*

****Lanius ludovicianus excubitorides** Swainson.--Miller, Friedmann, Griscom, and Moore (1957:216) recorded *L. l. excubitorides* from Sabinas and from Hipolito (November 2 to February 24).

***Lanius ludovicianus mexicanas** Brehm.--*Specimens examined:* total 4: sex ? 31089 from 7 mi. S, 2 mi. E Boquillas, 800 ft., February 29, 1952, weight, 45.1 gms.; 31090 from 36 mi. S, 15 mi. E Boquillas, 2550 ft., March 12, 1952; sex ? 30233 from 1 mi. N San Lorenzo, 4200 ft., February 5, 1951; and 32098 from 7 mi. S, 4 mi. E Bella Union, 7200 ft., June 25, 1952.

Miller (1931:66) suggested that *L. l. mexicanus* and *L. l. excubitorides* intergrade in Coahuila; all of the specimens of Loggerhead Shrike from Coahuila that I have examined are intergrades between *mexicanus* and *excubitorides.* Our four specimens have a superciliary line that is indistinct and the black mask of each extends somewhat posterior to the auricular region. The anterior part of their forehead is somewhat lighter than the remaining part of their head and back.

Miller (1955a:171) detected *L. l. mexicanus* only once in catclaw scrub in the lower part of Boquillas wash at about 4600 feet in the Sierra del Carmen. He (*loc. cit.*) remarked that his individual may best be considered an intergrade between *mexicanus* and *excubitorides,* being "somewhat closer to the former." Burleigh and Lowery (1942:201) obtained a male *L. l. mexicanus* "in the open valley just below the summit of Diamante Pass on April 23" that was typical of this subspecies. Burleigh and Lowery (1942:201-202) suggested that *L. l. mexicanus* breeds in southeastern Coahuila. Sutton and Burleigh

(1939a:39) took a typical representative *L. l. mexicanus* at Mayran on January 30.

Lanius ludovicianus gambeli Ridgway.--Miller, Friedmann, Griscom, and Moore (1957:214) recorded this subspecies from Hipolito on November 6.

****Vireo atricapilla*** Woodhouse.--*Specimens examined:* total 4: sex ? 32099-32100 from Sierra del Pino (=6 mi. N, 6 mi. W Acebuches), 5250 ft., July 3, 1952; and 31493-31494 from 16 mi. E, 18 mi. N Ocampo, May 8 and 9, 1954, enlarged testes.

The Black-capped Vireo seems to breed as far south as central Coahuila. Miller (1955a:171-172) reported this vireo as a summer resident in "the low catclaw-dominated scrub in the lower washes of Boquillas Canyon and its side valleys, at 4600 to 4800 feet" in the Sierra del Carmen. Dickerman found the Black-capped Vireo common on the dry scrub and oak hillside habitat 16 mi. E and 18 mi. N Ocampo.

Vireo griseus noveboracensis (Gmelin).--Ridgway (1904:184) recorded this subspecies of the White-eyed Vireo "west to Sabinas." Miller, Friedmann, Griscom, and Moore (1957:220) also recorded *V. g. noveboracensis* from Sabinas on May 25. I suspect that the specimen of *V. g. noveboracensis* from Sabinas that was taken on May 25 was a vagrant. *V. g. noveboracensis* may occur in Coahuila as a migrant or winter visitant; however, I do not believe that representatives of *noveboracensis* normally are resident in Coahuila.

****Vireo griseus micrus*** Nelson.--*Specimen examined:* one, 32101, from 2 mi. W Jimenez, 850 ft., June 20, 1952, measurements: wing, 58 mm.; tail, 43.5 mm.; culmen, 10 mm.; tarsus, 19 mm.

This subspecies of the White-eyed Vireo breeds in Coahuila. Miller, Friedmann, Griscom, and Moore (1957:221) recorded *V. g. micrus* from Sabinas on March 9 and May 14. The enlarged testes (5x3 mm.) of No. 32101 and the date (June 20) on which it was obtained suggest breeding in Coahuila.

Vireo huttoni stephensi Brewster.--This subspecies of Hutton's Vireo occurs in southeastern Coahuila as a migrant. Sutton and Burleigh (1939a:39) found *V. h. stephensi* fairly common at Diamante Pass on March 6. Ridgway (1904:198) recorded *V. h. stephensi* from Sierra Guadalupe in April.

****Vireo huttoni carolinae*** Brandt.--*Specimens examined:* total 3: 31588 from 20 mi. S Ocampo, 6500 ft., April 5, 1954, weight, 11 gms.; 32851 from 13 mi. E San Antonio de las Alazanas, July 6, 1955, weight, 14.2 gms.; and 32102 from 7 mi. S, 4 mi. E Bella Union, 7200 ft., June 25, 1952.

To my knowledge, *V. h. carolinae* is the only resident subspecies of Hutton's Vireo in Coahuila. Burleigh and Lowery (1942:202) found *V. h. carolinae* in rather limited numbers in the woods bordering the summit of Diamante Pass. Miller (1955a:172) remarked that the subspecies *carolinae* was a common bird from 6500 feet to 8000 feet in the Sierra del Carmen. Miller (*loc. cit.*) took a female on April 12 that was nearly ready to lay and said that his specimens of *carolinae* from the Sierra del Carmen seem to be separate from *V. h. stephensi* and *V. h. mexicanus.* Our specimens showed no overlapping of characters with *V. h. stephensi* and *V. h. mexicanus*. The size of the largest ovum (2 mm.) of No. 32102 and the dates (June 25, and July 6) on which Nos. 32102 and 32851 were obtained suggest that *V. h. carolinae* is a resident in Coahuila.

Vireo bellii medius* Oberholser.--*Specimen examined:* one, 31495, from San Marcos, May 5, 1954, measurements: wing, 56 mm.; tail, 48 mm.; culmen, 9.5 mm.; tarsus, 18 mm.

Although Bell's Vireo seems to have been observed uncommonly in Coahuila, this species does breed in the State. Burleigh and Lowery (1942:202) obtained a single female *V. b. medius* on April 20 "in an arroyo east of Saltillo" and found this subspecies "not uncommon in the open desert twenty miles west of Saltillo, where three singing males were secured." Miller, Friedmann, Griscom, and Moore (1957:226) recorded *V. b. medius* in breeding condition at Hipolito from June 30 to July 3. Ridgway (1904:207) recorded the subspecies from Monclova. Hardy saw a Bell's Vireo at Parras on July 4, 1955. The size of No. 31495 is typical for *V. b. medius.*

Vireo flavifrons* Vieillot.--*Specimen examined:* one, 32103, from 2 mi. W Jimenez, 850 ft., June 20, 1952.

Van Hoose (1955:302-303) suggested that the occurrence of No. 32103 in Coahuila is evidence of a southward extension of the range of the Yellowthroated Vireo within the last generation. The size of the testes (5x3 mm.) of No. 32103 and the date (June 20) on which it was obtained suggest breeding by *Vireo flavifrons* in Coahuila.

Vireo solitarius solitarius (Wilson).--*Specimen examined:* one, 31640, from 9 mi. E Hermanas (=Canyon de Parajos in the Sierra de Parajos Azule), 2100 ft., December 7, 1953.

This subspecies of the Solitary Vireo is an uncommon migrant or winter visitant to Coahuila. Dickerman obtained No. 31540, whose skull was unossified and whose ovary was small, in an oak and palm habitat. The bright yellow flanks, large and yellow wing bars, and the uniform olive green back indicate that this specimen is a typical representative of *V. s. solitarius*. This subspecies was previously unrecorded in Coahuila.

Vireo solitarius cassinii Xantus.--*Specimens examined:* total 2: 35408 (skeleton only) from 5 mi. W, 3 mi. S Acebuches, 6200 ft., May 12, 1954; and 31589 from 20 mi. S Ocampo, 6200 ft., April 4, 1954, measurements: wing, 75 mm.; tail, 57 mm.; culmen, 10.5 mm.; weight, 14 gms.

This subspecies of the Solitary Vireo seems to be uncommon in Coahuila. No. 31589 provides the first record of *V. s. cassinii* in Coahuila. Van Hoose has (1955:303) erroneously reported that Dickerman obtained No. 31589 on July 4, 1954; the correct date is April 4, 1954.

Although the size of No. 31589 is large for *V. s. cassinii*, the color (sides and flanks with less yellow, more olive; narrow white wing bars) resembles that of typical representatives of *cassinii*. The testes of No. 31589 were not enlarged. Dickerman suggested that the female from 5 mi. W and 3 mi. S Acebuches showed some resemblance to *V. s. plumbeus*. This is not to be unexpected since the subspecies *plumbeus* has been reported from the Chisos Mountains of Texas (Van Tyne and Sutton, 1937:82) and from northern Chihuahua (Miller, Friedmann, Griscom, and Moore, 1957:227).

****Vireo olivaceus*** (Linnaeus).--*Specimen examined:* one, 32104, from 12 mi. N, 12 mi. W Jimenez, 850 ft., June 19, 1952.

The Red-eyed Vireo has been uncommonly reported from eastern Coahuila. Miller, Friedmann, Griscom, and Moore (1957:229) recorded *V. olivaceus* from Sabinas on May 22 and from Las Vacas Creek on June 7 as late spring migrants. Hellmayr (1935:131) listed the Red-eyed Vireo from northern Coahuila. Findley saw a Red-eyed Vireo 2 mi. W Jimenez on June 19, 1952. The size of the testes (5x3 mm.) of No. 32104 and the date (June 19) on which it was obtained indicate that the Red-eyed Vireo possibly breeds in northeastern Coahuila; if so, this is the first breeding record of the Red-eyed Vireo in Coahuila.

Vireo gilvus gilvus (Vieillot).--This subspecies of the Warbling Vireo is an uncommon migrant in Coahuila. Miller, Friedmann, Griscom, and Moore (1957:232) recorded *V. g. gilvus* from 12 mi. W Saltillo on September 28.

Mniotilta varia (Linnaeus).--*Specimen examined:* one, 31662, from the north slope of Sierra Guadalupe (=11 mi. S, 7 mi. W General Cepeda), 7800 ft., April 20, 1953, weight, 10 gms., testes not enlarged.

The Black and White Warbler is an uncommon visitant or migrant in Coahuila. Miller (1955a:172) remarked that Marsh took a fall migrant on

September 1 in Chuperosa Canyon in the Sierra del Carmen. Burleigh and Lowery (1942:202) secured a female Black and White Warbler "in an orchard on the outskirts of Saltillo" on April 20 and a male "near the top of Diamante Pass on April 23."

Vermivora celata celata (Say).--*Specimens examined:* total 2: sex ? 31091 from the Rio Grande (=17 mi. S Dryden, Terrell Co., Texas, in Coahuila), 600 ft., March 19, 1952, measurements: wing, 57 mm.; tail, 47 mm.; weight, 7.7 gms.; and 31092 from 4 mi. W Hacienda La Mariposa, 2300 ft., March 25, 1952, measurements: wing, 62 mm.; tail, 48 mm.; weight, 9.2 gms.

This subspecies of the Orange-crowned Warbler is an uncommon migrant in Coahuila. In Brewster County, Texas, Van Tyne and Sutton (1937:83) found *V. c. celata* "not common as a spring transient." Miller, Friedmann, Griscom, and Moore (1957:239) recorded *V. c. celata* from Coahuila. The quality of the pale yellow color and the sizes of Nos. 31091-31092 suggest that they are representatives of *V. c. celata.*

Vermivora celata orestera Oberholser.--Burleigh and Lowery (1942:202) found *V. c. orestera* "only on infrequent occasions ... in the open woods surrounding the summit of Diamante Pass" where they obtained one specimen.

Vermivora virginiae (Baird).--Miller (1955a:172) took a male Virginia's Warbler in Boquillas Canyon in the Sierra del Carmen "in scattered scrubby oak growth with grass and cactus beneath." This species in the Sierra del Carmen is considered "casual" by Miller, Friedmann, Griscom, and Moore (1957:241).

****Vermivora crissalis** (Salvin and Godman).--*Specimen examined:* one, 31590, from 20 mi. S Ocampo, 7000 ft., April 5, 1954, weight, 10 gms.

The Colima Warbler is common locally in Coahuila. Burleigh and Lowery (1942:203) found this species fairly common on the steep, rugged slopes above the summit of Diamante Pass and saw none below an elevation of approximately 7500 feet. Bangs (1925:251) stated that Nelson and Goldman secured a specimen of the Colima Warbler at Sierra Guadalupe on April 25.

****Vermivora superciliosa mexicana*** (Bonaparte).--*Specimen examined:* one, 31591, from 13 mi. E San Antonio de las Alazanas, 9350 ft., April 10, 1954.

Hartlaub's Warbler is uncommon in Coahuila and seems to occur only in the southeastern section of the State; No. 31591 is the first record of the species in Coahuila. The size of the testes (5x2 mm.) of No. 31591 and the fact that the bird was singing when first seen suggest the possibility that

V. s. mexicana breeds in southeastern Coahuila. Breeding there is not unexpected because the species has been found breeding in Nuevo Leon (Miller, Friedmann, Griscom, and Moore, 1957:242).

Parula americana (Linnaeus).--Miller (1955a:172) obtained a migrant Parula Warbler in an oak grove at 7000 feet on April 16 in the Sierra del Carmen, and remarked that it was "apparently the first record of this species in Coahuila."

*****Parula pitiayumi nigrilora*** Coues.--The AOU Check-list Committee (1957:486) recorded this subspecies of the Olive-backed Warbler as a resident at Sabinas.

*****Peucedramus taeniatus arizonae*** Miller and Griscom.--This subspecies of the Olive Warbler is locally common in Coahuila. Miller (1955a:172) found *P. t. arizonae* common in the pine timber above 6800 feet in the Sierra del Carmen and suggested that this warbler breeds in these mountains. Burleigh and Lowery (1942:203) found the Olive Warbler in a thick pine wood at an elevation of 9500 to 10,000 feet and remarked that this species was "decidedly uncommon in the Diamante Pass area." They (*loc. cit.*) obtained a female (not identified to subspecies) on April 23 that had a "well developed brood patch and was unquestionably incubating eggs." Sutton and Burleigh (1939a:40) took a single female at Diamante Pass on March 6 which also was not identified to subspecies. Dickerman saw Olive Warblers in the Sierra de la Madera on December 13, 1953, and 13 mi. E San Antonio de las Alazanas on April 10, 1954.

Dendroica petechia morcomi Coale.--Miller, Friedmann, Griscom, and Moore (1957:246) reported this subspecies of the Yellow Warbler as having been recorded from Coahuila.

Dendroica auduboni auduboni (Townsend).--*Specimens examined:* total 2: 31094 (skeleton only) from Fortin (=33 mi. N, 1 mi. E San Geronimo), 3300 ft., March 29, 1952; and 31093 from 4 mi. W Hacienda La Mariposa, 2300 ft., March 25, 1952, weight, 12.3 gms.

Audubon's Warbler is a common winter visitant and migrant in Coahuila. Miller (1955a:173) recorded *D. a. auduboni* as a migrant from April 7 to 26 in the Sierra del Carmen; he found no suggestion of breeding by the Audubon's Warbler in the northwestern section of the State. One individual that Miller (*loc. cit.*) obtained was extensively black and approached the characters of the subspecies *nigrifrons* of Chihuahua. He (*loc. cit.*) suggested that the black individual was taken from "part of a cline of blackness and size in which *D. a. auduboni* of the northwest and *D. a. nigrifrons* of Mexico are extremes." Burleigh and

Lowery (1942:203) remarked that Audubon's Warbler "is doubtless a common winter bird in the area around Saltillo." Two specimens obtained by Burleigh and Lowery (*loc. cit.*) "might be considered intermediate" between *auduboni* and *memorabilis*. Sutton and Burleigh (1939a:40) saw Audubon's Warbler "in some numbers near San Pedro ... [on] January 29 and 30." Dickerman saw Audubon's Warblers 13 mi. E San Antonio de las Alazanas on April 10, 1954. Miller (1955a:173) also obtained, in the Sierra del Carmen, a hybrid between *D. coronata* and *D. auduboni*.

Dendroica auduboni memorabilis Oberholser.--Oberholser (1921:246) recorded *D. a. memorabilis* from Saltillo on April 17. This subspecies seems to winter commonly in western Mexico and less commonly in the Central Plateau and Sierra Madre Oriental (Miller, Friedmann, Griscom, and Moore, 1957:249-250).

Dendroica nigrescens (Townsend).--*Specimen examined:* one, 31095, from Fortin (=33 mi. N, 8 mi. W San Geronimo), 3300 ft., March 28, 1952, weight, 9.3 gms.

The Black-throated Gray Warbler is an uncommon spring and possibly fall migrant in Coahuila. Miller (1955a:173) found *D. nigrescens* uncommon in the Sierra del Carmen. He saw and heard a spring migrant singing on April 12 at 7000 feet and obtained a male on April 16.

Dendroica townsendi (Townsend).--Townsend's Warbler is a spring and fall migrant in Coahuila. Miller (1955a:173) recorded *D. townsendi* on September 2 at Jardin del Sur in Chuperosa Canyon in the Sierra del Carmen. Burleigh and Lowery (1942:203) saw two individuals of Townsend's Warbler at Diamante Pass on April 14. Amadon and Phillips (1947:578) secured this species "in mesquite about twenty miles west of Saltillo on August 28." Dickerman saw Townsend's Warblers in the Sierra de la Madera on December 13, 1953; 20 mi. S Ocampo on April 4, 1954; and 13 mi. E San Antonio de las Alazanas on April 10, 1954.

Dendroica virens (Gmelin).--Dickerman saw one Black-throated Green Warbler 13 mi. E San Antonio de las Alazanas, 9350 feet, on April 10, 1954, in a white pine-Douglas fir-aspen association. This seems to be the first record of this species in Coahuila.

__Dendroica chrysoparia__ Sclater and Salvin.--Miller, Friedmann, Griscom, and Moore (1957:251) listed the Golden-cheeked Warbler from Hipolito on July 3.

Dendroica occidentalis (Townsend).--*Specimen examined:* one, 31592, from 13 mi. E San Antonio de las Alazanas, April 10, 1954.

The Hermit Warbler seems to be an uncommon spring and probably fall migrant in Coahuila. Miller, Friedmann, Griscom, and Moore (1957:252) reported *D. occidentalis* from the State. No. 31592, whose testes were not enlarged, was obtained in a pine-spruce-aspen association.

Geothlypis trichas brachidactylus (Swainson).--Miller, Friedmann, Griscom, and Moore (1957:261) listed this subspecies of the Yellowthroat as recorded from Coahuila.

Geothlypis trichas occidentalis Brewster.--Burleigh and Lowery (1942:204) found the Yellowthroat "only on the outskirts of Saltillo, where ... [this species] was not uncommon in the thickets, especially around a small stand of marsh grass bordering a pond." Three males that Burleigh and Lowery (*loc. cit.*) obtained were tentatively identified as *G. t. occidentalis*. Dickerman saw one male Yellowthroat at San Marcos (=20 mi. S Cuatro Cienegas) on May 4, 1954.

Geothlypis nelsoni nelsoni Richmond.--Burleigh and Lowery (1942:204) noted the Hooded Yellowthroat "only on the open slopes above the summit of Diamante Pass at an elevation of about 8,000 feet" where they obtained an adult male.

*****Icteria virens virens*** (Linnaeus).--*Specimens examined:* total 3: 32105-32106 from 12 mi. N, 12 mi. W Jimenez, 850 ft., June 19, 1952; and 32107 from 2 mi. W Jimenez, 850 ft., June 20, 1952.

From the paucity of records in the literature, I judge that the Yellow-breasted Chat is uncommon in Coahuila. Miller, Friedmann, Griscom, and Moore (1957:265) listed *I. v. virens* from Coahuila. Findley saw a Yellow-breasted Chat 2 mi. S and 3 mi. E San Juan de Sabinas on June 22, 1952. The sizes of the testes (7x4 mm.; 12 mm.) of Nos. 32105-32106, the size of the largest ovum (2 mm.) of No. 32107, and the dates (June 19, 20) on which these specimens were obtained indicate breeding by *I. v. virens* in northeastern Coahuila, an area southwest of the previously documented breeding range.

*******Icteria virens auricollis*** (Deppe).--Miller, Friedmann, Griscom, and Moore (1957:264) listed this subspecies of the Yellow-breasted Chat as recorded from Coahuila.

Wilsonia pusilla pileolata (Pallas).--*Specimens examined:* total 3: 31501 and 31500 from Sierra del Pino (=5 mi. S, 3 mi. W Acebuches), May 13 and 14, 1954, measurements: wing, 59, 55 mm.; tail, 50, 49 mm.; culmen, 8, 8.5 mm.; tarsus, 16, 16 mm.; weight: 6, 7 gms.; and 31663 from the north foot of Sierra Guadalupe (=10 mi. S, 5 mi. W General Cepeda),

6500 ft., April 21, 1953, measurements: wing, 58 mm.; tail, 46 mm.; culmen, 8.5 mm.; tarsus, 16 mm.; weight, 7 gms.

Wilson's Warbler is a common spring and probably fall migrant in Coahuila. Miller (1955a:173) took spring migrants of *W. p. pileolata* from April 9 to April 27; he found *W. p. pileolata* at 4800 feet and at 7000 feet. Amadon and Phillips (1947:579) saw a Wilson's Warbler at Las Delicias on August 17. Burleigh and Lowery (1942:204) found Wilson's Warbler to be the most abundant of the warblers that they recorded in southeastern Coahuila. They saw *W. p. pileolata* on the top of the high ridges and in the arid desert country in the southeastern section of the State. Several specimens were collected by Burleigh and Lowery (*loc. cit.*) including an immature male from the Chorro del Agua on April 19. Dickerman saw Wilson's Warblers 16 mi. E and 18 mi. N Ocampo on May 7, 1954, and at San Marcos (=20 mi. S Cuatro Cienegas) on May 4, 1954. The sizes of our specimens as well as their color (bright olive-green above, bright yellow below) are typical for the subspecies *pileolata*.

Setophaga ruticilla ruticilla (Linnaeus).--The American Redstart seems to be uncommon in Coahuila. Miller, Friedmann, Griscom, and Moore (1957:268) listed one specimen of *S. r. ruticilla* from the State.

**Setophaga picta picta* Swainson.--*Specimens examined:* total 2: sex ? 31096 from 26 mi. W Santa Teresa, 7050 ft., April 5, 1952; and 31671 from Canon d. Meco in Sierra Guadalupe (=10 mi. S General Cepeda), April 23, 1953, weight, 11 gms.

The Painted Redstart seems to be locally common in Coahuila. Miller (1955a:173) found these warblers breeding in the Sierra del Carmen and said that they were "common from 6000 to 7500 feet in canyon bottom growth, in oaks, and in mixed pines and oaks;" however, he did not see any of these warblers above 7500 feet where conifers tended to dominate the vegetation. Marsh and Stevenson (1938:287) obtained a male Painted Redstart in annual molt on August 11 in oak and juniper forest at Vivoras Spring in the Sierra del Carmen and reported seeing two other Painted Redstarts at 9000 feet (see also Miller, 1955a:173). Dickerman also saw seven Painted Redstarts 20 mi. S Ocampo on April 4, 1954. The size of the testes (7x4 mm.) of No. 31671 suggests breeding by *S. p. picta* in the Sierra Guadalupe.

***Passer domesticus domesticus* (Linnaeus).--Burleigh and Lowery (1942:204) remarked that the House Sparrow was not "a common bird around Saltillo" although they noticed *P. d. domesticus* regularly. They reported House Sparrows also from the Chorro del Agua and in the high mountain valley south of Diamante Pass. Miller, Friedmann, Griscom, and Moore (1957:275) recorded *P. d. domesticus* from Sabinas. Baker captured House Sparrows in a bat net 12 mi. E San Antonio de las Alazanas, 9950 feet, on July 5, 1955. Although there are no other records, the House Sparrow is probably fairly common in the villages and towns of the State.

[****Sturnella magna hoopesi*** Stone.--The Eastern Meadowlark is uncommon in Coahuila. The AOU Check-list Committee (1957:523) listed this subspecies of the Eastern Meadowlark from northern Coahuila.]

****Sturnella neglecta neglecta*** Audubon.--*Specimen examined:* one, 31098, from the Rio Grande (=17 mi. S Dryden, Terrell Co., Texas, in Coahuila), 600 ft., March 18, 1952, weight, 71.8 gms.

This subspecies of the Western Meadowlark seems to be locally common in the open country of Coahuila. Burleigh and Lowery (1942:205) found this meadowlark common "about Saltillo" where a "small series" of *S. neglecta* was obtained. Miller, Friedmann, Griscom, and Moore (1957:295) recorded the Western Meadowlark from El Diamante on July 7. To my knowledge, no specific breeding records of this meadowlark from Coahuila exist.

****Xanthocephalus xanthocephalus*** (Bonaparte).--*Specimen examined:* one, 32494 (skeleton only) from Las Margaritas, August 4, 1955.

The Yellow-headed Blackbird occurs in Coahuila in migration. Miller (1955a:173) found this blackbird at Noria "in the flats immediately east of the Sierra del Carmen" on April 28, and reported also that Marsh took a male in worn breeding plumage on July 24 at Tanque de los Melones on La Bavia Ranch east of Fresno Mesa. Amadon and Phillips (1947:579) took two adult males at Las Delicias on August 11 and 15. Dickerman also saw a female 8 mi. E and 2 mi. S Americanos on May 18, 1954. Van Hoose saw a male at Torreon on July 2, 1955.

****Agelaius phoeniceus megapotamus*** Oberholser.--*Specimens examined:* total 5: 32124, 32126, and 32128 from 12 mi. N, 12 mi. W Jimenez, 850 ft., June 19, 1952; 32125 from 2 mi. W Jimenez, 850 ft., June 20, 1952; and 32127 from 9 mi. S, 11 mi. E Sabinas, June 14, 1952.

This subspecies of the Redwinged Blackbird is common in eastern Coahuila. There are no records of the species from western Coahuila. Burleigh and Lowery (1942:205) obtained a male at "the edge of Saltillo" on April 24. Oberholser (1919a:23) recorded *A. p. megapotamus* from Porfirio Diaz on June 2, 5, and 6. The presence of juveniles (32126, 32128) from 12 mi. N and 12 mi. W Jimenez and (32125) from 2 mi. W Jimenez, respectively, and the dates (June 14, 19, 20) on which the University of

Kansas specimens were obtained are evidence of breeding by *A. p. megapotamus* in northeastern Coahuila.

***Icterus spurius** (Linnaeus).--*Specimens examined:* total 8: 31536-31537 from 12 mi. N, 12 mi. W Jimenez, 850 ft., June 19, 1952; 31538 from 2 mi. S, 3 mi. E San Juan de Sabinas, 1160 ft., June 23, 1952; 32605-32607 from Parras, July 4, 1955; 32604 and 32603 from Hacienda San Lorenzo, July 3, 1955, weights, 19.4, 18.5 gms.

The Orchard Oriole seems to occur fairly commonly in eastern and southern Coahuila and breeds in the State. Amadon and Phillips (1947:579) reported that Orchard Orioles were common in the desert "about Las Delicias" in August and September and probably were migrants. Dickerman collected Nos. 32605-32606 along an irrigated field-edge that consisted of cottonwood and oak; he obtained Nos. 32603-32604 in an irrigated pecan orchard. The sizes of the testes (10x5 mm.; 10x5 mm.; 8x7 mm.; 8x7 mm.; 10x8 mm.) of Nos. 31536, 31537, 32605, and 32604, respectively, and the size of the largest ovum (2 mm.) of No. 32603 as well as the dates (June 19, 23; July 3, 4) on which all these specimens were collected indicate breeding by this species in the State.

***Icterus cucullatus cucullatus** Swainson.--*Specimens examined:* total 2: 32123 from 2 mi. W Jimenez, 850 ft., June 21, 1952; and 32122 from 2 mi. S, 3 mi. E San Juan de Sabinas, 1160 ft., June 23, 1952.

The Hooded Oriole apparently is uncommon in Coahuila. Miller, Friedmann, Griscom, and Moore (1957:290) listed *I. c. cucullatus* from Sabinas. The size of the testes (11x6 mm.) of No. 32122 and the dates (June 21, 23) on which Nos. 32123 and 32122 were collected as well as the juvenile male (tail, 80.5 mm.) from 2 mi. W Jimenez suggest breeding by this subspecies in Coahuila.

***Icterus parisorum** Bonaparte.--*Specimen examined:* one, 32121, from 7 mi. S, 4 mi. E Bella Union, 7200 ft., June 25, 1952.

Scott's Oriole is common in Coahuila. Miller (1955a:173) found this oriole breeding in the canyons at the base of the Sierra del Carmen. Burleigh and Lowery (1942:205) found this oriole limited to the higher altitudes above 7000 feet, and took specimens at Diamante Pass and at the Chorro del Agua on April 19. Amadon and Phillips (1947:579) found Scott's Oriole "not uncommon in the arroyos near Las Delicias" and reported a juvenile "barely out of the nest and able to fly only a few feet ..." on August 15. No. 32121 had an egg in its oviduct. Dickerman saw Scott's Orioles in the Sierra del Pino on May 12, 1954, and 16 mi. E and 18 mi. N Ocampo on May 7, 1954.

****Icterus wagleri wagleri** Sclater.--Ridgway (1902:268) recorded *Icterus wagleri* from Saltillo. Hellmayr (1937:122-123) referred this record of Wagler's Oriole to *I. w. wagleri.*

***Icterus bullockii bullockii** (Swainson).--This subspecies of Bullock's Oriole was listed as breeding by Miller, Friedmann, Griscom, and Moore (1957:282) at Monclova on May 12-19.

Euphagus cyanocephalus (Wagler).--Brewer's Blackbird is a common migrant in Coahuila. Miller (1955a:174) found a few as migrants in the Sierra del Carmen on April 27. Burleigh and Lowery (1942:205) remarked that "this blackbird was characteristically a bird of the towns and villages, the scattered flocks being invariably seen feeding in the streets and near the houses." They (*loc. cit.*) obtained three specimens at Diamante Valley on April 23 and remarked that the departure of these birds in spring was "extremely late."

***Cassidix mexicanus prosopidicola** Lowery.--*Specimens examined:* total 3: 32893 from Parras, July 4, 1955, weight, 98.8 gms.; and 35418-35419 from Torreon, January 8, 1954.

This subspecies of Boat-tailed Grackle has been recorded from several localities in Coahuila. In southeastern Coahuila, Burleigh and Lowery (1942:205-206) noted this grackle "somewhat local in its distribution; it was found in cultivated fields about the towns and villages." These authors noted it at Saltillo, the Chorro del Agua, and "occasionally in the open valley south of Diamante Pass" and obtained specimens from "near Saltillo" and Diamante Valley. On August 18 Amadon and Phillips (1947:579) found *C. m. prosopidicola* at Las Delicias where "a fledged young was noticed begging for food...." Lowery (1938:4) recorded one specimen of *C. m. prosopidicola* from Monclova. Findley saw Boat-tailed Grackles 2 mi. W Jimenez on June 19, 1952, and 2 mi. S and 11 mi. E Nava on June 15, 1952.

The distribution and intergradation of Boat-tailed Grackles in Coahuila is presently poorly understood. *C. m. prosopidicola* from southeastern Coahuila may approach *C. m. mexicanus,* and there is probable intergradation of *prosopidicola* with *monsoni* in northwestern Coahuila (Phillips, 1950:78).

Molothrus ater ater (Boddaert).--*Specimen examined:* one, 31513, from 21 mi. S, 11 mi. E Australia, 4400 ft., May 3, 1954, measurements: wing, 102 mm.; tail, 64 mm.; tarsus, 26 mm.; culmen, 17 mm.

This subspecies of the Brown-headed Cowbird is uncommon in Coahuila. Amadon and Phillips (1947:579) took an adult male and a juvenile female *M. a. ater* at Las Delicias on August 15, both of which were considered early migrants. Dickerman obtained No. 31513 from a flock of eight cowbirds. Although the measurements of this specimen agree equally well with those of *M. a. ater* and *M. a. artemisiae* (Grinnell, 1909:275-281), the shape of the bill most closely resembles that of *ater*. Grinnell (1909:278) said that "*ater* has a tumid bill, broad and high at [the] base with [a] conspicuously arched culmen" whereas "*artemisiae* has a longer and relatively much slenderer bill, vertically shallow at [the] base and laterally compressed, with the culmen in its greater portion straight or even slightly depressed." The size of the ovary (8x4 mm.) of No. 31513 and the date (May 3) on which it was obtained suggest that this individual was a late migrant, still south of the breeding range of *M. a. ater*.

Molothrus ater artemisiae Grinnell.--This subspecies of the Brown-headed Cowbird is an uncommon migrant in Coahuila. Amadon and Phillips (1947:579) obtained, at Las Delicias, a juvenile male on August 15 and an adult male on August 17.

*****Molothrus ater obscurus*** (Gmelin).--*Specimens examined:* total 18: 32112-32115 from 12 mi. N, 12 mi. W Jimenez, 850 ft., June 18 and 19, 1952; 32108-32111, 32116, and 32117-32120 from 2 mi. W Jimenez, 850 ft., June 20, 1952; 32491 from Las Margaritas, August 4, 1955; 31511 and 31510 from 16 mi. E, 18 mi. N Ocampo, May 8 and 7, 1952; and 35409-35410 (skeletons only) from 4 mi. N San Isidro, May 11, 1954.

This subspecies of the Brown-headed Cowbird is common in Coahuila and breeds there. Amadon and Phillips (1947:579) suggested that *M. a. obscurus* breeds at Las Delicias. Burleigh and Lowery (1942:206) saw the Brown-headed Cowbird in "small numbers on the outskirts of Saltillo...."

Measurements of the adult males that I have examined are: wing, 101.1 mm. (97-106); tail, 66.5 mm. (62-69); tarsus, 25.6 mm. (24.5-28); culmen, 17.3 mm. (16-18.5). Measurements of the adult females that I have examined are: wing, 92.3 mm. (90-97); tail, 60.1 mm. (56.5-62.5); tarsus, 23.5 mm. (22.5-24); culmen, 14.4 mm. (14-15). The sizes of the testes of three of the males (6-7 mm. long) and of the largest ova of four of the females (6-9 mm. in diameter) indicate breeding by this subspecies in Coahuila, as does the small size of one of the juvenile males (tail, 33.5 mm. long).

Piranga ludoviciana (Wilson).--In Coahuila the Western Tanager occurs fairly commonly as a migrant. There are no records of it breeding in the State. Miller (1955a:174) remarked that Marsh took a migrant Western Tanager at Jardin del Sur in the Sierra del Carmen on September 7. Amadon and Phillips (1947:579) took an adult male Western Tanager at Las Delicias on August 12. Miller, Friedmann, Griscom, and Moore (1957:305) also recorded this tanager from 12 mi. W Saltillo. Dickerman saw Western Tanagers in the Sierra del Pino on May 12, 1954, and 16 mi. E and 18 mi. N Ocampo on May 7, 1954.

***Piranga flava dextra** Bangs.--*Specimen examined:* one, 31526, from Sierra del Pino (=5 mi. S, 3 mi. W Acebuches), May 12, 1954, weight, 41 gms.

Miller (1955a:174) found this subspecies of the Hepatic Tanager "in the pine-oak belt at 7000 feet on April 12 [in the Sierra del Carmen], when a male was seen and a female taken." No. 31526 was with a female when taken; this male was not fat and its testes were not enlarged. The size of the wing (105.5 mm.) of No. 31526 represents the extreme maximum in this subspecies.

Piranga flava hepatica (Swainson).--Miller, Friedmann, Griscom, and Moore (1957:303) stated that *P. f. hepatica* is found in northwestern and central Arizona and southwestern New Mexico south into the highlands of Mexico, west of the Sierra Madre Oriental, to Oaxaca and in winter and migration extends eastward and south to Chiapas. These authors remarked also that *P. f. dextra* occurs in the mountains east of the continental divide in New Mexico and western Texas south through eastern Mexico to Chiapas. Specimens of the Hepatic Tanager from Coahuila in winter might well be either *P. f. hepatic* or *P. f. dextra*. Miller, Friedmann, Griscom, and Moore (*loc. cit.*) recorded migrants of *P. f. hepatica* from the Sierra de Guadalupe on April 24-27.

***Piranga rubra rubra** (Linnaeus).--*Specimens examined:* total 6: 32129, 32132, and 32133 from 12 mi. N, 12 mi. W Jimenez, 850 ft., June 18 and 19, 1952; 32130 from 2 mi. W Jimenez, 850 ft., June 20, 1952; and 32131 and 32134 from 2 mi. S, 11 mi. E Nava, June 15, 1952.

This Summer Tanager occurs in northeastern Coahuila. The specimens from 12 mi. N and 12 mi. W Jimenez, 2 mi. W Jimenez, and 2 mi. S and 11 mi. E Nava are typical representatives of *P. r. rubra*. The large testes (12 mm.) of No. 32129 and the well-developed brood patch of No. 32134 are evidence of breeding by this subspecies in the State. Heretofore this subspecies has not been recorded from Coahuila.

***Piranga rubra cooperi** Ridgway.--*Specimens examined:* total 4: 32828-32829, 32831, and 32830 from Parras, July 4, 1955.

This subspecies of Summer Tanager seems to occur throughout Coahuila except in the northeastern section of the State. Miller, Friedmann, Griscom, and Moore (1957:302) listed *P. r. cooperi* from Sabinas and Sierra de Guadalupe. Miller (1955a:174) saw a Summer

Tanager in a canyon in the Serranias de Burros, about 40 miles east of the Sierra del Carmen on April 28. Although there are no other records of *P. r. cooperi* from northwestern Coahuila, Van Tyne and Sutton (1937:96) recorded this tanager as a common nesting species in Brewster County, Texas, in cottonwood, mesquite, or willow trees. I suspect that *P. r. cooperi* is a common nesting bird in northwestern Coahuila as well.

Nos. 32828-32831 approach *P. r. rubra*. The measurements of No. 32829 are: wing, 98 mm.; tail, 80 mm.; the measurements of No. 32831 are: wing, 98 mm.; tail, 79 mm. The specimens of *P. r. cooperi* from Parras are somewhat small and seemingly approach *P. r. rubra*. The sizes of the testes (8x5 mm.; 9x5 mm.) of Nos. 32829 and 32831, respectively, and the size of the largest ovum (4x4 mm.) of No. 32830 indicate breeding by this subspecies in southern Coahuila, as does the presence of No. 32828, a juvenile male.

***Richmondena cardinalis canicaudus** (Chapman).--*Specimens examined:* total 3: 31099 from 1 mi. N Boquillas, 700 ft., March 8, 1952, weight, 45.3 gms.; 32135 from 2 mi. W Jimenez, 850 ft., June 20, 1952; and 32136 from 2 mi. S, 3 mi. E San Juan de Sabinas, June 22, 1952.

In Coahuila the Cardinal is common. Miller (1955a:174) found it singing in the Boquillas drainage of the Sierra del Carmen at 4800 feet, and gave evidence that the Cardinal breeds there. Sutton and Burleigh (1939a:43) found the species to be common in the low country "east of Saltillo." Hellmayr (1938:69) recorded *R. c. canicaudus* from Sabinas. The sizes of the testes (9 mm.; 7x4 mm.) of Nos. 32135-32136 indicate breeding by this subspecies in northeastern Coahuila.

***Pyrrhuloxia sinuata sinuata** (Bonaparte).--*Specimens examined:* total 4: 31100 from 10 mi. S, 5 mi. E Boquillas, 1500 ft., March 4, 1952, weight, 37.3 gms.; 32137 from 5 mi. N, 19 mi. W Cuatro Cienegas, 3250 ft., July 5, 1952; 35403 (skeleton only) from San Marcos, May 5, 1954; and 30234 from 3 mi. SE Torreon, 3800 ft., January 12, 1951.

In Coahuila, the Pyrrhuloxia is common. Hellmayr (1938:76) listed it from Sabinas. Ridgway (1901:628) recorded *P. s. texana* (=*sinuata*) from La Ventura. Burleigh and Lowery (1942:206) found the species "only in the open desert country west of Saltillo where, on April 22, several pairs were seen in a small arroyo." Amadon and Phillips (1947:579) took an immature *P. s. sinuata* at Las Delicias; Sutton and Burleigh (1939a:43-44) found this subspecies fairly common in the San Pedro district on January 29 and 30. Miller, Friedmann, Griscom, and Moore (1957:330) recorded breeding by *P. s. sinuata* at Hipolito on July 2. The size of the testes (8 mm.) of No. 32137 indicates breeding in central Coahuila.

****Pheucticus melanocephalus melanocephalus*** (Swainson).--*Specimen examined:* one, 31664, from Sierra Guadalupe (=10 mi. S, 5 mi. W General Cepeda), 7500 ft., April 30, 1953, measurements: wing, 105.5 mm.; tail, 78 mm.; culmen, 19 mm.; weight, 48 gms.

Miller (1955a:174) reported that the Black-headed Grosbeak first appeared in the Sierra del Carmen on April 13 and was soon seen patrolling territories. He remarked that these specimens from the Sierra del Carmen conformed adequately with the rather poorly differentiated race *P. m. melanocephalus* and stated that Marsh took an immature male at Jardin del Sur on September 7. Oberholser (1919b:416) listed *Hedymeles melanocephalus papago* (=*P. m. melanocephalus*) from Sierra Guadalupe on April 27. Dickerman saw Black-headed Grosbeaks in the Sierra del Pino on May 12, 1954, and 16 mi. E and 18 mi. N Ocampo on May 7, 1954. The size of No. 31664 represents the characters of *P. m. melanocephalus* as presented by Oberholser (1919b:413). No specimen of *P. m. melanocephalus* from Coahuila, to my knowledge, approaches *P. m. maculatus*. The size of the testes (7x5 mm.) of No. 31664 suggests breeding by this subspecies in the Sierra Guadalupe.

****Guiraca caerulea interfusa*** Dwight and Griscom.--*Specimens examined:* total 2: 32138 from 12 mi. N, 12 mi. W Jimenez, 850 ft., June 18, 1952; and 32139 from 2 mi. W Jimenez, 850 ft., June 21, 1952.

Burleigh and Lowery (1942:206) recorded this Blue Grosbeak from "about twenty miles west of Saltillo" on April 22. Miller (1955a:174) stated that Marsh obtained a male *G. c. interfusa* at Vivoras Spring on August 3. Miller, Friedmann, Griscom, and Moore (1957:334) remarked that the Blue Grosbeak breeds in the northern and eastern sections and reported *G. c. interfusa* from Hipolito on July 2. Amadon and Phillips (1947:580) reported examining material from Sabinas referable to *G. c. interfusa*. Dickerman saw Blue Grosbeaks 4 mi. N San Isidro on May 10, 1954. Findley saw Blue Grosbeaks 4 mi. W Jimenez on June 19, 1952, and 2 mi. S and 3 mi. E San Juan de Sabinas on June 22, 1952. Nos. 32138-32139 are typical representatives of *G. c. interfusa*. The size of the testes (12 mm.) of No. 32138, the size of the largest ovum (2 mm.) of No. 32139, and the dates (June 18, 21) on which they were collected are evidence of breeding by this subspecies.

*****Guiraca caerulea eurhyncha*** Coues.--Amadon and Phillips (1947:580) obtained an adult male of this Blue Grosbeak from Las Delicias on August 12. This subspecies, according to Miller, Friedmann, Griscom, and Moore (1957:334), is resident at middle and lower elevations through most of central and southern Mexico. Except for the occurrence of intergrades of *G. c. interfusa, caerulea,* and *eurhyncha* in southern Nuevo Leon and Tamaulipas (Miller, Friedmann, Griscom, and Moore, 1957:335), the record from Las Delicias represents the northern limit of the range of the subspecies *eurhyncha.*

****Passerina cyanea*** (Linnaeus).--*Specimen examined:* one, 31544, from San Marcos, May 5, 1954.

The Indigo Bunting is rare in Coahuila. Van Hoose (1955:303) reported that No. 31544 seems to provide the first record of the species in the State. The Indigo Bunting is a summer resident in southwestern Oklahoma and southeastern Texas (Miller, Friedmann, Griscom, and Moore, 1957:336). No. 31544 seemingly represents a large extension in the summer range of the Indigo Bunting. Van Hoose (*loc. cit.*) stated that No. 31544 was observed chasing another male, and the pursuer in turn was followed by a female; he thought that the territorial behavior of this bird suggested breeding.

****Passerina versicolor versicolor*** (Bonaparte).--*Specimens examined:* total 2: 35414 from 6 mi. N, 2 mi. W Castillon, 3750 ft., June 30, 1953, weight, 8 gms.; and 35415 from 5 mi. S Castillon, 4150 ft., June 28, 1953.

Although the Varied Bunting has been recorded only from northwestern Coahuila, I suspect that this bird is locally common throughout most of the State. Miller (1955a:174) stated that the habitat of this species consisted of catclaw-covered bottom lands at the base of the Sierra del Carmen at 4700 feet. Miller's records indicate incipient breeding by *P. v. versicolor* in the Sierra del Carmen on April 26. The size of the testes (11 mm.) of No. 35415 and the dates (June 28, 30) on which Nos. 35414-35415 were collected are strong evidence of breeding by the Varied Bunting 6 mi. N and 2 mi. W Castillon and 5 mi. S Castillon.

****Passerina ciris pallidior*** Mearns.--*Specimens examined:* total 6: 32141 and 32142 from 2 mi. S, 11 mi. E Nava, 810 ft., June 15 and 16, 1952; 35416 from 6 mi. N, 2 mi. W Castillon, 3750 ft., June 29, 1953, weight, 15 gms.; 32140 from 2 mi. S, 3 mi. E San Juan de Sabinas, June 22, 1952; 31547 from 16 mi. N, 14 mi. E Ocampo, May 10, 1954, weight, 16 gms.; and 31546 from San Marcos, May 5, 1954.

The Painted Bunting is a common summer resident in Coahuila. Marsh and Stevenson (1938:287) wrote that Painted Buntings were common in summer in the foothills of the Sierra del Carmen, and they took a male at Piedra Blanca on July 25 and a female at Jardin del Sur on September 1. Amadon and Phillips (1947:580) remarked that two immature Painted Buntings, "apparently migrants, were taken at Las

Delicias on August 10 and 11." Miller, Friedmann, Griscom, and Moore (1957:339) recorded breeding by *P. c. pallidior* at Hipolito, 4200 feet, on July 1. Findley saw Painted Buntings 2 mi. S and 11 mi. E Nava on June 15, 1952. Dickerman saw Painted Buntings 4 mi. N San Isidro on May 10, 1954. Hardy saw Painted Buntings at Parras on July 4, 1955.

All the University of Kansas specimens are typical of the larger subspecies *pallidior*. The dates (May 5, 10; June 15, 16, 22, 29) on which these specimens were collected, the sizes of the testes (9x6 mm.; 6x5 mm.; 7x3 mm.) of Nos. 32140, 31547, and 31546, respectively, and the size of the largest ovum (2.5 mm.) of No. 32142 indicate breeding by *P. c. pallidior* in Coahuila.

Carpodacus cassinii Baird.--Cassin's Finch is an uncommon winter migrant in Coahuila. Miller, Friedmann, Griscom, and Moore (1957:314) listed *C. cassinii* from Sierra Guadalupe.

****Carpodacus mexicanus potosinus*** Griscom.--*Specimens examined:* total 8: 35417 from 13 mi. S, 5 mi. W Castillon, 4000 ft., June 28, 1953, testes, 6 mm., weight, 19 gms.; 35411 (skeleton only) from Sierra del Pino (=5 mi. W, 3 mi. S Acebuches), 6200 ft., May 14, 1954; 32145 from 5 mi. N, 19 mi. W Cuatro Cienegas, 3250 ft., July 5, 1952; 32144 and 32147 from 3 mi. S, 3 mi. E Bella Union, 6750 ft., June 27, 1952, gonads, 7 mm.; 32143 and 32146 from 7 mi. S, 4 mi. E Bella Union, 7200 ft., June 24, 1952; and 31675 from the west foot of Pico de Jimulco, 5000 ft., April 4, 1953, weight, 20 gms.

The House Finch is common in Coahuila. Miller (1955a:174) reported a sparse population of this species in the foothills adjoining Boquillas Canyon at 4800 feet of the Sierra del Carmen and observed young just out of the nest on April 25. He remarked also that specimens of the House Finch from the Sierra del Carmen seem to show no intergradation toward *frontalis*. Burleigh and Lowery (1942:206) noted *C. m. potosinus* at "Saltillo, in the desert country west of there, at the Chorro del Agua, and in the open valley south of Diamante Pass." Miller, Friedmann, Griscom, and Moore (1957:316) listed breeding by *C. m. potosinus* at El Diamante on July 6. Dickerman also saw the House Finch at San Marcos on May 4, 1954, and Hardy saw it at Parras on July 4, 1955. No. 32147 had a distinct brood patch; the largest ovum of No. 32146 was 7 mm. in diameter. No. 32145 was a juvenile male.

The University of Kansas specimens agree well with descriptions of *C. m. potosinus* as given by Moore (1939:195). No approach toward *C. m. frontalis, centralis,* or *nigrescens* is exhibited by any of these specimens. No. 31675, from Pico de Jimulco in southwestern Coahuila, is paler above and below than any other specimens of *C. m. potosinus*. Also the crown of No. 31675 is suffused with more red than in typical representatives of *C. m. potosinus*.

****Spinus pinus pinus** (Wilson).--*Specimen examined:* one, sex ? 33219 (skeleton only) from 13 mi. E San Antonio de las Alazanas, 9950 ft., July 6, 1955.

On April 5, 7, and 21, Miller (1955a:175) recorded winter visitant or vagrant flocks of Pine Siskins in the Sierra del Carmen. Miller, Friedmann, Griscom, and Moore (1957:318) listed *S. p. pinus* south to Sierra Guadalupe. Dickerman saw Pine Siskins 13 mi. E San Antonio de las Alazanas on April 10, 1954.

Spinus pinus macropterus (Bonaparte).--Miller, Friedmann, Griscom, and Moore (1957:319) recorded a vagrant *S. p. macropterus* from 50 mi. S Monclova, 2850 feet, on November 9.

Spinus tristis pallidus Mearns.--*Specimen examined:* one, 31101, from Fortin (=33 mi. N, 8 mi. W San Geronimo), 3300 ft., March 28, 1952, weight, 11.5 gms.

The American Goldfinch is an uncommon migrant or winter visitant in Coahuila. Hellmayr (1938:296) recorded *S. t. pallidus* from Sabinas. Fortin and Sabinas are the only places in Coahuila where *S. t. pallidus* has been collected. No. 31101 is a typical representative of *S. t. pallidus*.

****Spinus psaltria psaltria** (Say).--*Specimens examined:* total 7: 32148-32149 and 32151 from 12 mi. N, 12 mi. W Jimenez, 850 ft., June 19, 1952; 32150 from 2 mi. W Jimenez, June 20, 1952; 33220 (skeleton only) from Parras, July 4, 1955; and 32939 and 32940 from Mesa de las Tablas, July 7, 1955, weights, 9.5, 11 gms.

In Coahuila, the Lesser Goldfinch seems to be common. Although Miller (1955a:175) did not find it in the Sierra del Carmen, he reported that Marsh took a specimen on August 22 in Chuperosa Canyon that was "presumed to" be *S. p. psaltria*. Burleigh and Lowery (1942:206) observed the Lesser Goldfinch "on the outskirts of Saltillo in an orchard on April 20."

Nos. 32148-32151 and 32940 were typical for the subspecies *psaltria*. A partial albino (32939), which was obtained from a pine-oak-wheat field edge, has upper parts that lack the black coloring of typical representatives of *S. p. psaltria*. Instead the crown and back of No. 32939 is yellow, resembling the color of its underparts, the wing coverts are white, and its primaries are black with white edgings.

*__Atlapetes pileatus dilutus__ Ridgway.--*Specimens examined:* total 2: 32942 and 32943 from 13 mi. E San Antonio de las Alazanas, 9950 ft., July 6, 1955.

The Rufous-capped Atlapetes occurs uncommonly in southeastern Coahuila. The male and female *A. p. dilutus* from 13 mi. E San Antonio de las Alazanas represent the first records of occurrence of this subspecies in the State. The smaller size, grayer upper parts, and pale yellow color of the underparts of Nos. 32942-32943 characterize the subspecies *dilutus*. The size of the testes (7x6 mm.) of No. 32942 and the date (July 6) on which both specimens were collected indicate breeding by this species in Coahuila.

*__Arremonops rufivirgata rufivirgata__ (Lawrence).--*Specimens examined:* total 2: 32152 and 32153 from 2 mi. S, 3 mi. E San Juan de Sabinas, June 22 and 23, 1952.

The Olive Sparrow is uncommon in Coahuila. Miller, Friedmann, Griscom, and Moore (1957:349) listed *A. r. rufivirgata* from Sabinas on February 10. Sabinas and southeast of San Juan de Sabinas seem to be the only localities in Coahuila where the Olive Sparrow has been collected and also are at the westernmost extremity of range of this species. The size of the testes (9x7 mm.) of No. 32152 and the dates of collection indicate breeding by the Olive Sparrow in Coahuila.

Chlorura chlorura (Audubon).--*Specimen examined:* one, 30238, from 1 mi. N San Lorenzo, 4200 ft., February 5, 1951.

The Green-tailed Towhee is a common migrant and winter visitant in Coahuila; the species has been found at several localities. Miller (1955a:175) noted several migrants "each day in the last week of April ... at the mouth of Boquillas Canyon" of the Sierra del Carmen. Burleigh and Lowery (1942:207) noted the species "in an arroyo in the open desert country about twenty miles west of Saltillo on April 22." Miller, Friedmann, Griscom, and Moore (1957:351) listed the Green-tailed Towhee from 12 mi. W Saltillo on September 28 and from Sabinas.

*__Piplio erythrophthalmus gaigei__ Van Tyne and Sutton.--*Specimens examined:* total 3: 31102 from Fortin (=33 mi. N, 8 mi. W San Geronimo), 3300 ft., March 28, 1952, weight, 38.8 gms.; 35412 (skeleton only) from Sierra del Pino (=5 mi. W, 3 mi. S Acebuches), 6200 ft., May 15, 1954, testes enlarged, weight, 34 gms.; and 31593 from 17 mi. S Ocampo, 5300 ft., April 7, 1954, weight, 38 gms.

The Rufous-sided Towhee is locally common in Coahuila; *P. e. gaigei* is present in northern Coahuila. Miller (1955a:175) remarked that "between 6800 and 7500 feet these towhees were sparsely distributed in areas of scattered low ceanothus and hawthorne, chiefly in canyon bottoms, but also on slopes where ceanothus was intermingled with downed timber and young pines." He (*loc. cit.*) indicated that the adult obtained by Marsh from Vivoras Canyon on August 25 was *P. e. gaigei* rather than *P. maculatus montanus* (see also Sibley, 1950:127). Dickerman saw Rufous-sided Towhees in the Sierra de la Madera on December 13, 1953. I have referred Nos. 31102 and 31593 to *gaigei* although both are close to *orientalis*. The size of the testes (14x8 mm.) of No. 31593 suggests breeding.

***Piplio erythrophthalmus orientalis** Sibley.--*Specimens examined:* total 3: 32154 from 7 mi. S, 4 mi. E Bella Union, 7200 ft., June 25, 1952; 33223 (skeleton only) from 13 mi. E San Antonio de las Alazanas, 9950 ft., July 6, 1955; and 31630 from Mesa de Tablas, 8600 ft., January 15, 1954, weight, 46 gms.

This subspecies of the Rufous-sided Towhee occurs in southeastern Coahuila. Burleigh and Lowery (1942:207) stated that it "was limited in its distribution to the mountain sides ... [and was] noted in the arroyos at the base of the mountains and from there up to about 8,000 feet." These authors identified their specimens from Diamante Pass and from Saltillo as *Pipilo maculatus gaigei*. Sibley (1950:129) reidentified them, as well as a series from Sierra de Guadalupe, as *P. e. orientalis*. The size of the testes (12x7 mm.) of No. 32154 as well as the date (June 25) on which it was obtained suggests breeding by the Rufous-sided Towhee in southeastern Coahuila.

***Pipilo fuscus potosinus** Ridgway.--*Specimens examined:* total 3: 32155 from 7 mi. S, 4 mi. E Bella Union, 7200 ft., June 25, 1952; 31676 from the west foot of Pico de Jimulco, 5000 ft., April 5, 1953, weight, 45 gms.; and sex ? 29560 from 7 mi. S, 1 mi. E Gomez Farias, 6500 ft., November 19, 1949, weight, 46.5 gms.

The subspecies *potosinus* has been recorded from several localities in Coahuila. Davis (1951:70) listed the following localities in the State from which *P. f. potosinus* has been collected: Muralla, San Lazaro Mountains, 50 mi. S Monclova, 2850 ft., Saltillo; Saltillo (Chorro de Agua); 19 mi. W Saltillo; Cresta Blanca, 12 mi. W Saltillo, 5500 ft.; Diamante Pass, 11 mi. S Saltillo, 6000-8000 ft.; and Carneros. Miller, Friedmann, Griscom, and Moore (1957:358) recorded a "small juvenile" from El Diamante on July 5. Sutton and Burleigh (1939a:45) recorded *P. f. texanus* from Diamante Pass on March 6. I suspect that Davis (*op. cit.*) reidentified the specimen concerned from Diamante Pass as *P. f. potosinus*. Burleigh and Lowery (1942:207) indicated that *P. f. potosinus* "was fairly common and of general distribution in the area, occurring both on the arid plateau about Saltillo and on the mountain sides up to an elevation of about 8,000 feet." Nos. 32155, 31676, and 29560 are typical for *P. f. potosinus*. The size of the testes (14x7 mm.) of No. 32155 suggests breeding by the Brown Towhee in southeastern Coahuila.

****Pipilo fuscus texanus*** van Rossem.--*Specimen examined:* one, 31103, from 10 mi. S, 5 mi. E Boquillas, 1500 ft., March 5, 1952, weight, 46.9 gms.

This subspecies of the Brown Towhee occurs in northwestern Coahuila south through the Sierra del Carmen. Miller (1955a:176) reported that his series of Brown Towhees from the Sierra del Carmen agreed satisfactorily with *texanus* although revealing some sign of intergradation with *potosinus*. Davis (1951:70) thought that *P. f. potosinus* is present in northern as well as southern Coahuila. Miller (1955a:176), however, remarked that *P. f. texanus* is more characteristic of the population of Brown Towhees of northwestern Coahuila. He indicated that a single juvenile taken by Marsh on August 28 from Jardin del Sur and allocated to *P. f. potosinus* by Davis probably is *P. f. texanus*.

Calamospiza melanocorys Stejneger.--*Specimen examined:* one, 30239, from 10 mi. E Torreon, 3700 ft., January 9, 1951.

The Lark Bunting is an uncommon winter visitant in Coahuila. Other than the present specimen, the only record of the Lark Bunting in Coahuila is that of Burleigh and Lowery (1942:207), who reported a small flock of this species from which several specimens were collected "on April 20 in a field on the edge of Saltillo."

Passerculus sandwichensis oblitus Peters and Griscom.--Miller, Friedmann, Griscom, and Moore (1957:364) recorded *P. s. oblitus* from Sabinas on February 25 and March 18 and 24. These records represent sparse winter visitants to Coahuila.

Passerculus sandwichensis brooksi Bishop.--Burleigh and Lowery (1942:208) recorded *P. s. brooksi* from Diamante Pass in April; Miller, Friedmann, Griscom, and Moore (1957:363) remarked that this specimen is perhaps best regarded as a variant of one of the races normally wintering in this area.

Passerculus sandwichensis anthinus Bonaparte.--*Specimens examined:* total 3: 31104 from 10 mi. E Hacienda La Mariposa, 2000 ft., March 30, 1952, weight 15 gms.; 31105 from 8 mi. N, 4 mi. W Muzquiz, 1800 ft., April 1, 1952, weight, 19.0 gms.; and 31594 from 17 mi. S Ocampo, 5300 ft., April 7, 1954, weight, 16 gms.

Burleigh and Lowery (1942:208) recorded *P. s. anthinus* from Diamante Pass. This subspecies is not uncommon in Coahuila. The University of Kansas specimens showed no indication of breeding.

Passerculus sandwichensis nevadensis Grinnell.--This subspecies of the Savannah Sparrow is uncommon in Coahuila; Hellmayr (1938:490) listed one specimen from Sabinas.

Passerculus sandwichensis brunnescens (Butler).--Burleigh and Lowery (1942:208) recorded this subspecies of the Savannah Sparrow from Diamante Pass in April.

Ammodramus savannarum perpallidus (Coues).--*Specimen examined:* one, 31562, from 3 mi. N, 4 mi. E San Francisco (=25 mi. N Ocampo), 4850 ft., May 16, 1954, weight, 15 gms.

The Grasshopper Sparrow is an uncommon spring and possibly fall migrant in Coahuila. Miller, Friedmann, Griscom, and Moore (1957:367) recorded *A. s. perpallidus* from Sabinas on March 12. No. 31562, which was obtained in a yucca and acacia association, had little fat.

Ammodramus bairdii (Audubon).--Miller, Friedmann, Griscom, and Moore (1957:368) remarked that Baird's Sparrow is a rare winter visitant to the northern states of Mexico and recorded *A. bairdii* from Saltillo on May 8.

Pooecetes gramineus confinis Baird.--The Vesper Sparrow seems to be an uncommon winter visitant in Coahuila. Miller (1955a:176) found *P. g. confinis* "on two occasions in the grass of the dry cienega at the head of Corte Madera Canyon at 7500 feet" on April 9 and 14 in the Sierra del Carmen. In April, Burleigh and Lowery (1942:208) found *P. g. confinis* only in Diamante Valley where this sparrow "appeared to be quite uncommon." Sutton and Burleigh (1939a:45) took a male *P. g. confinis* at San Pedro on January 29.

*****Chondestes grammacus strigatus*** Swainson.--*Specimen examined:* one, 32156, from 8 mi. N, 2 mi. W Piedras Negras, June 18, 1952.

The Lark Sparrow is not uncommon in Coahuila. Miller (1955a:176) saw one male, "apparently on a breeding territory, on April 27 in an open, rather barren desert flat adjoining the lower part of Boquillas wash at 4600 feet." He reported that Marsh took a young of the year, still largely in juvenile plumage, on September 6 in the Sierra del Carmen. Amadon and Phillips (1947:580) remarked that Lark Sparrows were common "about Las Delicias" after August 18. Findley saw Lark Sparrows 2 mi. W Jimenez on June 19, 1952, and 2 mi. S and 11 mi. E Nava on June 15, 1952. Dickerman saw Lark Sparrows at San Marcos on May 4, 1954. The pale and narrowly streaked upperparts of No. 32156 are typical for *C. g. strigatus*. The size of the testes (9x4 mm.) of No. 32156 and the date (June 18) on which it was obtained suggest breeding by the Lark Sparrow in northeastern Coahuila.

*****Aimophila ruficeps tenuirostris*** Burleigh and Lowery.--This subspecies of the Rufous-crowned Sparrow is resident in the northern part of Coahuila. Miller (1955a:176) remarked that the species "ranged up to 7000 feet on open south-facing slopes within the oak belt" of the Sierra del Carmen. Specimens collected by him showed no approach to *boucardi* of southern Mexico and seem to be closest to *tenuirostris*. Miller referred the specimen that Marsh and Stevenson (1938:287) took on August 22 in Chuperosa Canyon to *tenuirostris* rather than *boucardi*. Miller, Friedmann, Griscom, and Moore (1957:376) listed *A. r. tenuirostris* from 50 mi. S Monclova on November 8 and 10.

*****Aimophila ruficeps boucardi*** (Sclater).--This subspecies of the Rufous-crowned Sparrow is common in southern Coahuila. Burleigh and Lowery (1942:208) indicated that *A. r. boucardi* was common throughout the mountainous areas and to some extent in the arroyos of the open desert country of southeastern Coahuila. Sutton and Burleigh (1939a:46) took a female *A. r. boucardi* "near Diamante Pass." Ridgway (1901:252) listed *A. r. boucardi* from Carneros. *A. r. tenuirostris* and *boucardi* seem to intergrade in central and even southern Coahuila. Miller, Friedmann, Griscom, and Moore (1957:376) recorded intermediate populations of the Rufous-crowned Sparrow from 12 mi. W Saltillo.

*****Aimophila cassinii*** (Woodhouse).--*Specimens examined:* total 2: 32157 from 9 mi. S, 11 mi. E Sabinas, June 14, 1952; and 32158 from 18 mi. S, 14 mi. E Tanque Alvarez, 4000 ft., July 6, 1952.

Cassin's Sparrow seems to be common in Coahuila. The AOU Check-list Committee (1957:603) listed *A. cassinii* from 10 mi. E Saltillo. Miller,

Friedmann, Griscom, and Moore (1957:378-379) recorded Cassin's Sparrow from Sabinas, on April 25; from 25 mi. SW Monclova, on November 20; from 12 mi. W Saltillo, on September 28; and from 10 mi. NE Saltillo, on July 3. These authors stated that Cassin's Sparrow was breeding 10 mi. NE Saltillo. The sizes of the testes (5x3 mm.; 7x5 mm.) of Nos. 32157-32158, respectively, and the dates (June 14, July 6) on which they were obtained are additional evidence of breeding by Cassin's Sparrow in Coahuila.

Amphispiza bilineata bilineata (Cassin).--*Specimens examined:* total 5: 32159 and 32163 from 2 mi. W Jimenez, 850 ft., June 20, 1952; 32160 from 8 mi. N, 2 mi. W Piedras Negras, June 18, 1952; 32161 and sex ? 32162 from 5 mi. N, 19 mi. W Cuatro Cienegas, 3250 ft., July 5, 1952.

The Black-throated Sparrow is common in Coahuila. Typical representatives of *A. b. bilineata* occur in eastern Coahuila. The center of the State is occupied by intergrades between *A. b. bilineata, opuntia,* and *grisea*. Miller, Friedmann, Griscom, and Moore (1957:381) reported *A. b. bilineata* from the "Saltillo area." Hellmayr (1938:539) recorded *A. b. bilineata* from Sabinas. The sizes (13, 14 mm.) of the white spot on the lateral tail feathers of Nos. 32163 and 32160, respectively, suggest *A. b. bilineata*. The short wing (61-64.5 mm.) and the lighter and browner color of the backs of Nos. 32163, 32160, and 32161 are suggestive of *A. b. bilineata*. Nevertheless, the size (8 mm.) of the white spot on the lateral tail feather of No. 32161 indicates intergradation with *A. b. opuntia*.

The sizes of the testes (6x4 mm.; 8 mm.) of Nos. 32160 and 32161, the size of the largest ovum (2 mm. in diameter) of No. 32163, and the presence of the juveniles from 2 mi. W Jimenez and 5 mi. N and 19 mi. W Cuatro Cienegas indicate breeding by *A. b. bilineata* in Coahuila.

Amphispiza bilineata opuntia Burleigh and Lowery.--*Specimens examined:* total 2: 31106 from 10 mi. S, 5 mi. E Boquillas, 1500 ft., March 3, 1952; weight, 12.3 gms.; and 31108 from 35 mi. S, 14 mi. E Boquillas, 2350 ft., March 12, 1952.

This subspecies of the Black-throated Sparrow occurs in northwestern Coahuila. Miller (1955a:176) stated that the Black-throated Sparrow was moderately common in the open desert scrub at the base of the Sierra del Carmen below 4800 feet. He said that the series of Black-throated Sparrows from the Sierra del Carmen "resembles most the race *opuntia* of western Texas ... but shows some intergradation toward *grisea* of southern Coahuila and toward *A. b. bilineata* of eastern Coahuila." He remarked also that the specimen from Jardin del Sur, which Marsh and Stevenson (1938:287) reported as *A. b. grisea*, was in extremely worn, dirty summer plumage and contributed nothing reliable to racial determination.

*__Amphispiza bilineata grisea__ Nelson.--*Specimen examined:* one, 31665, from the north foot of Sierra Guadalupe (=10 mi. S, 5 mi. W General Cepeda), 6500 ft., April 25, 1953.

A. b. grisea is the subspecies of Black-throated Sparrow in southern Coahuila. Burleigh and Lowery (1942:208) saw this sparrow "frequently on the arid plateau around Saltillo" and obtained specimens there that were identified as *A. b. grisea*. Amadon and Phillips (1947:581) saw individuals on August 8 and 28 that were feeding "fledged young near Saltillo." The size of its wing (68.5 mm.), the slaty color of its back, and the size (8 mm.) of the white spot on its lateral tail feather suggest that No. 31665 is characteristic of *A. b. grisea*. The size of the testes (4.5x3 mm.) of No. 31665 indicates that *A. b. grisea* may breed in southern Coahuila.

*__Junco phaeonotus palliatus__ Ridgway.--*Specimens examined:* total 3: 35402 (skeleton only) from 13 mi. E San Antonio de las Alazanas, 9345 ft., April 10, 1954; 33226 (skeleton only) from 13 mi. E San Antonio de las Alazanas, 9950 ft., July 6, 1955; and 31633 from Mesa de Tablas, 8600 ft., January 16, 1954, weight, 22 gms.

In Coahuila the Mexican Junco seems to be common. Miller (1955a:177) found it in the conifers of the upper Corte Madera drainage at 7500 feet and up to 8800 feet on Loomis Peak in the Sierra del Carmen. Marsh and Stevenson (1938:287) took an adult in Vivoras Canyon on August 14 in the Sierra del Carmen. Sutton and Burleigh (1939a:46) found small flocks at Diamante Pass in March. Burleigh and Lowery (1942:208-209) noted *J. p. palliatus* as a common bird of the mountain slopes above an elevation of about 7000 feet in southeastern Coahuila, and obtained specimens at Diamante Pass. Ridgway (1901:300) recorded *J. p. phaeonotus* from Sierra Encarnacion. His record should be of *J. p. palliatus* as indicated by Miller, Friedmann, Griscom, and Moore (1957:386). Dickerman saw Mexican Juncos in the Sierra de la Madera on December 13, 1953.

No. 31633 shows no sign of intergrading with *J. p. phaeonotus* to the south. The outermost rectrix of No. 31633 is wholly white; the second rectrix is nearly as white. No. 31633 is paler than representatives of *J. p. phaeonotus* from the southern part of the Central Plateau of Mexico.

Spizella passerina arizonae Coues.--*Specimens examined:* total 4: 31110 from the Rio Grande (=17 mi. S Dryden, Terrell Co., Texas, in Coahuila), 600 ft., March 18, 1952, weight, 10.7 gms. 31111 and 31112 from 4 mi. W Hacienda La Mariposa, 2300 ft., March 24, 1952, weights, 11.0, 11.7 gms.; and 31666 from the north foot of Sierra Guadalupe (=10 mi. S, 5 mi. W General Cepeda), 6400 ft., April 19, 1953, weight, 14.0 gms.

The Chipping Sparrow is a common spring and possibly fall migrant in Coahuila. Miller (1955a:177) noted small flocks from April 21 to 27 in the Sierra del Carmen where specimens were taken. Burleigh and Lowery (1942:209) indicated that *S. p. arizonae* was "quite common and of general distribution" at Saltillo and Diamante Pass and remarked that specimens were taken at these localities in April. Amadon and Phillips (1947:581) took two Chipping Sparrows "near Las Delicias on August 17." Dickerman saw individuals in the Sierra del Pino on May 12, 1954, and at San Marcos on May 4, 1954. Our specimens, which are typical representatives of *S. p. arizonae,* are pale; the ground color of their backs is grayish buff.

Spizella pallida (Swainson).--The Clay-colored Sparrow is a migrant or winter visitant in Coahuila. Burleigh and Lowery (1942:209) saw "large flocks of sparrows, mostly of this species, ... on frequent occasions in the cultivated fields and orchards on the outskirts of Saltillo." Four specimens were taken by Burleigh and Lowery (*loc. cit.*) on April 19 and 20 at Saltillo.

Spizella breweri breweri Cassin.--*Specimens examined:* total 2: 31115 from the Rio Grande (=17 mi. S Dryden, Terrell Co., Texas, in Coahuila), 600 ft., March 18, 1952, weight, 8.2 gms.; and 31114 from 28 mi. S, 11 mi. E Boquillas, 2000 ft., March 12, 1952, weight, 9.7 gms.

Brewer's Sparrow is probably a winter resident in much of Coahuila. Miller, Friedmann, Griscom, and Moore (1957:389) recorded *S. b. breweri* from 25 mi. NW Monclova on November 20 and from 8 mi. S Cuatro Cienegas on November 15. The only definite records obtained by Burleigh and Lowery (1942:209) of *S. b. breweri* are those of a female and a male taken "near Saltillo" on April 16 and 18. The size of the dorsal area of sandy buff with narrow streakings of Nos. 31114-31115 suggests *S. b. breweri.*

Spizella pusilla arenacea Chadbourne.--*Specimen examined:* one, 31116, from 4 mi. W Hacienda La Mariposa, 2300 ft., March 24, 1952, weight, 13.1 gms.

The Field Sparrow is an uncommon spring and probably fall migrant in Coahuila. Other than No. 31116, *S. p. arenacea* has only been recorded from Sabinas in March, when three specimens were obtained (Miller, Friedmann, Griscom, and Moore, 1957:390). The coloration of No. 31116 is much grayer and the black streaks on its back are much narrower than on typical representatives of *S. p. pusilla.*

****Spizella wortheni wortheni** Ridgway.--The single specimen of Worthen's Sparrow obtained by Burleigh and Lowery (1942:209) "just outside the limits of Saltillo on April 16" represents the only record of occurrence of this species in Coahuila.

****Spizella atrogularis atrogularis** (Cabanis).--Miller, Friedmann, Griscom, and Moore (1957:391) remarked that the Black-chinned Sparrow is a common resident of the Central Plateau from Durango and southern Coahuila southward. Burleigh and Lowery (1942:212) noted the species only at the foot of the mountains of southeastern Coahuila where, at 6000 feet, scattered pairs were found. Miller (1955a:177) observed a male on April 23 on a slope at the mouth of Boquillas Canyon of the Sierra del Carmen; he presumed it to be a transient.

Zonotrichia leucophrys leucophrys (Forster).--*Specimen examined:* one, 30243, from 1 mi. SW San Pedro de las Colonias, 3700 ft., February 8, 1951.

The White-crowned Sparrow is a fairly common migrant or winter visitant in Coahuila. Miller, Friedman, Griscom, and Moore (1957:393) recorded *Z. l. leucophrys* from Sabinas on March 16 and February 23. Burleigh and Lowery (1942:212) remarked that "small flocks of White-crowned Sparrows were seen at infrequent intervals in thickets and stretches of underbrush on the outskirts of Saltillo on April 18 and again on April 24 ... [and] near a small town some twenty miles west of Saltillo on April 22." Specimens that Burleigh and Lowery (*loc. cit.*) collected "near Saltillo" were identified as *Z. l. leucophrys.*

Zonotrichia leucophrys gambelii (Nuttall).--*Specimen examined:* one, 31117, from Sierra de la Encantada (=38 mi. S, 23 mi. E Boquillas), 4400 ft., March 15, 1952.

Miller (1955a:177) noted *Z. l. gambelii* on April 27 at 4600 feet in Boquillas wash in the Sierra del Carmen. Hellmayr (1938:568) listed *Z. l. gambelii* from Sabinas.

Zontrichia leucophrys oriantha Oberholser.--This subspecies has been recorded from Sabinas on April 25 and Hipolito on November 5 (Miller, Friedmann, Griscom, and Moore, 1957:393).

Melospiza lincolnii lincolnii (Audubon).--*Specimen examined:* one, 31595, from 20 mi. S Ocampo, 7000 ft., April 5, 1954, measurements: wing, 58 mm.; tail, 50 mm.; weight, 12 gms.

Lincoln's Sparrow seems to be a fairly common migrant or winter visitant in Coahuila; *M. l. lincolnii* is the common subspecies. Burleigh and Lowery (1942:212) found this sparrow only in a grain field situated between a small pond and a narrow stream on the outskirts of Saltillo; the four specimens collected were identified as *M. l. lincolnii*. No. 31595 was obtained in a pine-oak association.

Melospiza lincolnii alticola (Miller and McCabe).--Miller, Friedmann, Griscom, and Moore (1957:398) listed *M. l. alticola*, which seems to be uncommon in Coahuila, from Sabinas on March 14.

Melospiza lincolnii gracilis (Kittlitz).--Miller (1955a:177) took an unsexed representative of *M. l. gracilis* on April 7 in Carboneras Canyon of the Sierra del Carmen at 6700 feet. This occurrence is at the extreme eastern range of this subspecies.

Melospiza georgiana ericrypta Oberholser.--In Coahuila this subspecies of the Swamp Sparrow has been recorded as a migrant or winter visitant. Miller, Friedmann, Griscom, and Moore (1957:399) recorded it from Sabinas on February 22 to March 8 and from 8 mi. S Cuatro Cienegas on November 4.

THE END

Made in United States
Cleveland, OH
06 November 2026

47769828R00047